Cellular and Molecular Aspects of Ageing

Valquiria Bueno
Editor

Cellular and Molecular Aspects of Ageing

Springer

Editor
Valquiria Bueno
Federal University of São Paulo (UNIFESP)
São Paulo, Brazil

ISBN 978-3-031-55021-8 ISBN 978-3-031-55022-5 (eBook)
https://doi.org/10.1007/978-3-031-55022-5

This Springer imprint is published by the registered company Springer Nature Switzerland AG
The registered company address is: Gewerbestrasse 11, 6330 Cham, Switzerland

Paper in this product is recyclable.

Contents

Chapter 1
Introduction

Alphonse Laya and Valquiria Bueno

Cellular Senescence

Cellular senescence can be defined as a state of stable, terminal cell cycle arrest which is associated with various macromolecular changes and a hypersecretory, pro-inflammatory phenotype occurring during ageing [1]. Ageing people accumulate senescent cells that contribute to body dysfunction. This complex ubiquitous process has roles to wound repair and embryogenesis however in some cases this can lead to cancer development and age-related diseases [2]. Cellular senescence is characterized by a permanent cell cycle arrest in response to several insults. The main characteristics of senescent cells include changes in morphology, upregulation of cyclin-dependent kinase inhibitors, resistance to apoptosis, mitochondrial alterations, SASP (cytokines, chemokines, matrix remodeling proteins, growth factors). The presence of senescent cells in tissues has been suggested to play a role in age-related chronic diseases. In addition, cellular senescence has been proposed as a barrier in tumorigenesis but factors secreted by senescent cells (SASP) are involved in tumor progression (cancer metastasis and invasion). Markers of senescent cells are important as they can be targeted for the elimination of senescent cells (senolytics) or to inhibit SASP (senomorphics). Senolytics include the target of anti-apoptotic Bcl-2 (ABT23 and ABT737) or other apoptotic pathways, nuclear exclusion of p53 ($FOXO_4$-DRI) and glycolysis (quercetin, 2-DG) whereas senomorphics target members of SASP (i.e. NF-κB).

A. Laya
Department of Biological Sciences, Faculty of Science, University of Maroua, Maroua, Cameroon

V. Bueno (✉)
Department of Microbiology Immunology and Parasitology, UNIFESP Federal University of São Paulo, Botucatu St 862, 4o Level, São Paulo, SP 04023-0900, Brazil
e-mail: vbueno@unifesp.br

V. Bueno (ed.), *Cellular and Molecular Aspects of Ageing*,
https://doi.org/10.1007/978-3-031-55022-5_1

Genomic Instability

The complete and accurate DNA replication is important for the integrity of the transmission of genetic information in any organism. The chromosome instability is the common form of genomic instability [3]. Genomic instability can be caused by many factors including external and intracellular damage [4] due to reactive oxygen species which treats the DNA and mutation as well. Its consequences are age-related diseases including cancer and neurodegenerative disorders. Thus, cancer treatment based on blocking the genome instability may help patients' outcome. For the stability of genome, it is crucial an error-free replication of genetic material (DNA/RNA) and the repair of mistakes during replication or damaged DNA/RNA. DNA double-strand break (DSB) is repaired via homologous recombination (HR) and non-homologous end-join (NHEJ) which are essential to avoid chromosomal rearrangements that are deleterious or to prevent mutations. Changes in the mechanisms that regulates HR and DSR repair are linked to genomic instability and can cause mutations, aberrant chromosomal rearrangements, loss of heterozygosity of deleterious alleles. During the ageing process there is a diminished intrinsic efficiency of DNA repair pathways. In humans, the senescence of fibroblasts has been associated with decrease of efficiency and fidelity of DSB repair. In addition, in human cells the approaching of senescence present accumulation of chromosomal DSB. Culture of cells from patients with syndromes of premature ageing also presents cumulative DSB. As an example, it has been shown that the expression of laminin A and its mutated form (progerin) affects DSB repair and genomic stability. The increased expression of progerin has been associated with the Hutching-Gilford Progeria Syndrome (premature ageing syndrome) whereas in healthy individuals the low expression of progerin seems to play a role in the process of normal ageing.

Telomeres

Telomeres (protective caps of chromosome) are repetitive sequences at the ends of chromosomes that shorten over time due to insufficient replication of the 3′ end [5] which can induce senescence or apoptosis in order to protect the integrity of chromosome. Telomere is maintained through the enzyme telomerase which preserve the genomic integrity. Telomerase exerts critical extra-telomeric function in ageing through some proteins including telomerase reverse transcriptase (TERT) and RNA components, telomerase RNA component (*TERC*) [6]. The telomere shortening is mainly caused by the oxidative stress when the antioxidant compounds are in deficit or not working due to diseases. In this line, it was reported that vitamin C can restore telomere length [7]. In addition, the vitamin D supplementation demonstrated positive effect of telomere length in observational studies [8]. The telomeres attrition is associated with many age-related disorders including cancers and neurodegenerative diseases. In addition, critical short telomeres induce senescence in rapid dividing cells. Telomeres play an important role in cell fate and ageing since they protect the chromosome end from degradation and illegitimate recombination. Nevertheless, with cell division, telomeres undergo progressive shortening. After a critical length is reached, the telomeric DNA cannot form a protective secondary structure and cells enter in replicative senescence. Embryonic cell lines keep telomeres constant due to activated telomerase whereas in adult stem cells this activity compensates only partially for telomere shortening and in somatic cells it is not active. Dysfunctional telomeres activate a DNA damage response leading to cells with a senescence phenotype (decreased ability to proliferate and secretion of pro-inflammatory factors). Epidemiological studies have shown association between telomere attrition, ageing, and morbidity/mortality. The understanding of the binomial telomere shrink—cell senescence could lead to preventive and therapeutic interventions for age-related diseases.

Epigenetic

Epigenetic can be defined as the study of the modifications of DNA that affect the transcription of genes. This machinery may contribute to gene regulation in eukaryotic species [9]. Epigenetic alterations to DNA and chromatin are powerful regulators of genome architecture and gene expression, and they play a crucial role in mediating the induction and maintenance of senescence [10]. Alterations in the habits including diet and exercise represent an innovative epigenetic target to prevent and treat human diseases [11]. As an example, it was reported that dietary restrictions with good nutritional content can extend lifespan [11]. Plant metabolites with bioactive polyphenolics can inhibit DNA methyltransferase involved in preventing cancer, metabolic diseases and other chronic illnesses. In addition, it was reported that the epigenetic modifications, contribute to the progression of chronic diseases [12]. Epigenetic modifications observed on type 2 diabetes are DNA methylation, acetylation, ubiquitylation, SUMOylation, and phosphorylation at the lysine residue at the amino terminus of histones, affecting DNA histones, and non-coding RNA [13]. The DNA covering histone proteins forms chromatin which in association with epigenetic factors affects gene expression. These factors are enzymes modifying DNA directly or via histones (H2A, H2B, H3, H4, H3.3, H2AZ). Ageing has been associated with general loss of histones and local/global chromatin remodeling. In addition, it has been shown global and local change in DNA methylation, changes in heterochromatin, and reorganization of nuclei. The transcriptional program is changed in which senescent cells upregulate alternative histone transcripts and alter epigenetic by inducing inflammatory response genes (senescence-associated secretory phenotype). The link between epigenetic changes and age-related diseases can lead to targets for therapies aiming the improvement of healthspan.

Proteostasis

Proteostasis is a dynamic process in which mammalian cells can keep the balance of proteome by regulating protein synthesis, folding, transport, post-translational modification, and degradation [14]. The decline of protein homeostasis (proteostasis) is a hallmark in aging and for several age-related diseases [15]. Proteostasis is controlled by the proteostasis network, including chaperones, ubiquitin–proteasome and lysosomal-autophagy systems, which maintain the stability and function of the cellular proteome through preventing the accumulation of aggregated or damaged proteins [16]. Thus, ageing results in a decline of cellular proteostasis capacity which culminates in the accumulation of toxic material, causing the onset of age-related diseases and ultimately cell death [17]. Ageing has been related to reduced capacity of protein homeostasis (proteostasis) network causing thus cell dysfunction and contributing for the older adults increased susceptibility to chronic diseases. This network is responsible for the control of protein synthesis, folding and degradation. In neurodegenerative syndromes, the loss of efficacy in the proteostasis network leads to protein aggregation which is highly prevalent in the ageing population. The components of this system are: ribosome and associated factors (protein synthesis regulation), chaperones (folding and refolding of misfolding proteins), protein trafficking pathways for protein location, lysosome-autophagy system, ubiquitin–proteasome system (degrades misfolding proteins). Ageing has been linked to changes in the activity of chaperones/partner proteins which can compromise the folding of proteins. In addition, the reduced amount of ATP can affect the activity of chaperones and lead to the accumulation of damaged proteins.

Deregulated Nutrient-Sensing

Biological processes occurring in the organism requires nutrients and thus, the detection of nutrient availability via signaling pathways are crucial for tissue maintenance. Nutrient-sensing systems such as insulin/insulin-like growth factor 1 (IGF-1), mammalian target of rapamycin (mTOR), and 5'adenosine monophosphate-activated protein kinase (AMPK) have been shown to play a role in lifespan of *C. elegans*, *D. melanogaster*, and mice. IGF-1 has been linked to metabolism, protein homeostasis and responses to stress. mTOR regulates growth and metabolism in response to signals from IGF-1, AMPK, cytokines, amino acids, and oxygen levels. The inhibitor of mTORC1 (rapamycin) has been associated with increased lifespan in experimental models. AMPK senses the energy status linked to energy depletion in cells leading to catabolism and repressing anabolism. Nutrient-sensing systems interact not only in the intracellular, but also in the intercellular and intertissue levels and thus control energy storage or expenditure, somatic maintenance and lifespan. Clinical trials evaluating metformin, rapamycin and biochemical/clinical markers of ageing) have been reported.

Mitochondria/Oxidative Stress

Oxidative stress is the result of imbalance between the production of free radicals and the scavenging antioxidant systems in organisms [18]. The oxidative stress results in disruption of redox signaling and control, causing molecular damage [19]. In this process, mitochondria is the main organelle damaged. The mitochondrial genome or mitochondrial DNA (mtDNA) is released as a response to cellular stress. Mitochondria are essential organelles that can regulate cellular homeostasis via bioenergetic changes, reactive oxygen species (ROS) generation, signal transduction, and apoptosis. For instance, in ovarian aging and in type 2 diabetes, the oxidative stress (OS) is considered to be a key factor [20]. In addition, the antioxidant diets may be promising to reduce oxidative stress by decreasing oxidant emission and improving antioxidant mechanisms [19].

Ageing has been related to accumulation of errors in cells (nuclei and mitochondria) and damage in organelles and macromolecules. The process of mitophagy removes damaged mitochondria preventing thus immune activation via damage-associated molecular patterns (DAMPs). It has been shown that mitophagy declines during the ageing process which could lead to functional deficits and increase in the oxidative stress present in tissues and organs of ageing individuals. Mitochondrial dysfunction during the ageing process has been linked to functional declines such as sarcopenia, insulin resistance and neurological dysfunctions. In this context, signaling pathways maybe affected by lifestyle and pharmacologic components leading to reduction of age-related diseases.

Stem Cell Exhaustion

During replication cycle, stem cells generates stem or differentiated daughter cells. Embryonic stem cells (ESC) derive from blastocyst and being undifferentiated pluripotent cells, are responsive for a diversity of cell fates. In the postnatal organism, adult stem cells provide the differentiation into limited cell types. Tissue homeostasis and regeneration is dependent on adult stem cells amount and adequate function. In this context, stem cell exhaustion could be represented by the impairment in number and function of stem cell during the ageing process. Muscle, bone, and forebrain are maintained by adult stem cells and stem cell exhaustion reported in these tissues have been linked to age-related diseases. In addition, hematopoietic stem cell differentiation changes with age which affect the generation/maturation of immune cells. It has been suggested that the metabolic-epigenetic regulatory axis affects adults stem cells and pharmacological interventions such as metformin (AMPK), resveratrol (SIRT1), senolytics (elimination of senescent cells) have been proposed.

Another option is the use of induced pluripotent stem cells (iPSCs) which are obtained by forcing the nucleus of an adult cell into a pattern of altered gene expression. iPSCs are ESC-like somatic cells and both of them are considered as pluripotent stem cells due to self-renew and differentiation into a variety of cell types. iPSCs can be used for regenerative processes in patient-specific diseases and for the development of new drugs.

Cell to Cell Communication

The communication between cells plays a crucial role in physiological and pathological conditions. In this line, multipotent mesenchymal stromal cells (MSCs) use many mechanisms to communicate with cells of hematopoietic origin [21]. It was reported that MSCs are recognized to generate a wide range of exosomes (EXOs) (MSC-EXOs), showing that MSCs may be effective for EXO generation in a clinically appropriate measure as compared to other cell origins [22]. Exosomes are extracellular vesicles (EVs) including microvesicles, or apoptotic vesicles, and they play an essential role in intercellular communication [23]. The secretion of soluble factors by cells which affect other cells in the neighborhood is the best understood way of cell to cell communication. Senescent-associated secretory phenotype (SASP) is one of the ways of cell to cell communication and has been suggested as a target for intervention in age-related diseases. Classical SASP has been associated with the release of chemokines, cytokines, growth factors, and matrix remodeling enzymes. Other forms of communication include extracellular vesicles and receptor/ligand such as interleukin-1 receptor (IL-1R)/interleukin-1 alpha (IL-1A). Considering that cell to cell communication has been linked to benefits or negative effects for tissue homeostasis, the further knowledge in this field is crucial for therapies development and healthy ageing.

Senescence-Associated Secretory Phenotype (SASP)

Cellular senescence has been related to ageing mainly via senescence-associated secretory phenotype (SASP) and thus, the reduction of SASP abundance and consequently its action is a possible way to improve ageing conditions (i.e. cardiovascular disease, osteoporosis, diabetes, physical decline, brain dysfunction). SASP include interleukins, chemokines, growth factors, components of extracellular matrix, and proteases. In addition, lipid metabolites and extracellular vesicles have been recognized as novel SASP categories [24]. Senescent cells can also secrete extracellular vesicles (microRNA, protein, chromatin fragments from the cytoplasm) that affect tissue microenvironment (proliferation and inflammation). The main mediator of SASP at transcriptional level is NF-κB whereas at post-transcriptional level SASP is regulated by mTOR. Post-translational changes in SASP have been associated with the inflammasome and activation of IL-1-signaling pathway. Compounds such as rapamycin, resveratrol and metformin have been tested as reducers of SASP. However, these compounds are not specific for senescent cells and they are anti-inflammatory drugs that exert toxicities after continuous use. The further knowledge of SASP and its networks may lead to more specific compounds to promote healthspan.

Acknowledgements CAPES PrInt UNIFESP 88881.310735/2018-01.

References

1. Schmitt CA, Wang B and Demaria M (2022) Senescence and cancer — role and therapeutic opportunities. NAture revIeWS | CliniCAl OnCOlOgy 19, 619–636. https://doi.org/10.1038/s41571-022-00668-4.
2. Huang W, Hickson L T J, Alfonso Eirin A, Kirkland J L and Lerman LO (2022) Cellular senescence: the good, the bad and the unknown. NAture reviewS | NEPhRology. 18, 611–627.https://doi.org/10.1038/s41581-022-00601-z
3. Garribba L, De Feudis G, Martis V, Galli M, Marie Dumont M, Yonatan Eliezer Y, René Wardenaar R, et al. and Santaguida S. Short-term molecular consequences of chromosome mis-segregation for genome stability Nature Communications| (2023) 14:1353. https://doi.org/10.1038/s41467-023-37095-7.
4. Guo S, Zhu X, Huang Z, Wei C, Yu J, Zhang L, Feng J, Li M, Li Z (2023) Genomic instability drives tumorigenesis and metastasis and its implications for cancer therapy. Biomedicine & Pharmacotherapy 157 (2023) 114036. https://doi.org/10.1016/j.biopha.2022.114036.
5. Carlund O, Norberg A, Osterman P, Landfors M, Degerman S and Hultdin M (2023). DNA methylation variations and epigenetic aging in telomere biology disorders. Scientific Reports. 13:7955 | https://doi.org/10.1038/s41598-023-34922-1.
6. Denham J (2023) Canonical and extra-telomeric functions of telomerase: Implications for healthy ageing conferred by endurance training. 22, (6) e13836. https://doi.org/10.1111/acel.13836.
7. Cai Y, Zhong Y-d, Zhang H, Lu P-l, Liang Y-y, Hu B and Wu H (2023) Association between dietary vitamin C and telomere length: A cross-sectional study. Front Nutr 10:1025936. https://doi.org/10.3389/fnut.2023.1025936.
8. Rahman, S.T., Waterhouse, M., Pham, H. et al (2023) Effects of Vitamin D Supplementation on Telomere Length: An Analysis of Data from the Randomised Controlled D-Health Trial. J Nutr Health Aging. https://doi.org/10.1007/s12603-023-1948-3
9. Wang J, Shi A, and Lyu J (2023) A comprehensive atlas of epigenetic regulators reveals tissue-specific epigenetic regulation patterns. EPIGENETICS. 18 (1) 2139067. https://doi.org/10.1080/15592294.2022.2139067.
10. Crouch, J.; Shvedova, M.; Thanapaul, R.J.R.S.; Botchkarev, V.; Roh, D (2022) Epigenetic Regulation of Cellular Senescence. Cells 2022, 11, 672. https://doi.org/10.3390/cells11040672.
11. Abraham, M.J.; El Sherbini, A.; El-Diasty, M.; Askari, S.; Szewczuk, M.R (2023) Restoring Epigenetic Reprogramming with Diet and Exercise to Improve Health-Related Metabolic Diseases. Biomolecules. 13, 318. https://doi.org/10.3390/biom13020318.
12. Li D, Li Y, Yang S, Lu J, Jin X, Wu M (2022) Diet-gut microbiota-epigenetics in metabolic diseases: From mechanisms to therapeutics. Biomedicine & Pharmacotherapy 153 (2022) 113290. https://doi.org/10.1016/j.biopha.2022.113290.
13. Suárez, R.; Chapela, S.P.; Álvarez-Córdova, L.; BautistaValarezo, E.; Sarmiento-Andrade, Y.; Verde, L.; Frias-Toral, E.; Sarno, G. Epigenetics in Obesity and Diabetes Mellitus: New Insights. Nutrients 2023, 15, 811. https://doi.org/10.3390/nu15040811.

14. Tang J-X and Xiao F-H (2023) Editorial: The regulation of proteostasis in aging. Front. Cell Dev. Biol. 11:1221510. https://doi.org/10.3389/fcell.2023.1221510.
15. Neves D, Duarte-Pereira S, Matos S, Silva M R (2023) Proteostasis networks in aging: novel insights from text-mining approaches. Biogerontology (2023) 24:555–562. https://doi.org/10.1007/s10522-023-10027-0.
16. Liang, S., Li, D.-Y., Wen, J.-H., Tang, J.-X., and Liu, H.-F. (2022). Protein synthesis errors and longevity: A lesson from a single amino acid mutation study. Aging Dis. 13 (1), 1–3. https://doi.org/10.14336/AD.2021.1211.
17. van Oosten-Hawle P (2023), Exploiting inter-tissue stress signaling mechanisms to preserve organismal proteostasis during aging. Front. Physiol. 14:1228490. https://doi.org/10.3389/fphys.2023.1228490.
18. Zhang Z, Huang Q, Zhao D, Lian F, Li X and Qi W (2023) The impact of oxidative stress-induced mitochondrial dysfunction on diabetic microvascular complications. Front. Endocrinol. 14:1112363. https://doi.org/10.3389/fendo.2023.1112363.
19. Verhaegen, D.; Smits, K.; Osório, N.; Caseiro, A (2022) Oxidative Stress in Relation to Aging and Exercise. Encyclopedia 2022, 2, 1545–1558. https://doi.org/10.3390/encyclopedia2030105.
20. Yan F, Zhao Q, Li Y, Zheng Z, Kong X, Shu C, Liu Y and Shi Y (2022) The role of oxidative stress in ovarian aging: a review. Journal of Ovarian Research. 15:100. https://doi.org/10.1186/s13048-022-01032-x
21. Deniz, I.A., Karbanová, J., Wobus, M. et al (2023) Mesenchymal stromal cell-associated migrasomes: a new source of chemoattractant for cells of hematopoietic origin. Cell Commun Signal **21**, 36. https://doi.org/10.1186/s12964-022-01028-6.
22. Oveili, E., Vafaei, S., Bazavar, H. et al (2023) The potential use of mesenchymal stem cells-derived exosomes as microRNAs delivery systems in different diseases. Cell Commun Signal **21**, 20. https://doi.org/10.1186/s12964-022-01017-9.
23. Liu, YJ., Wang, C (2023) A review of the regulatory mechanisms of extracellular vesicles-mediated intercellular communication. Cell Commun Signal **21**, 77. https://doi.org/10.1186/s12964-023-01103-6.
24. Yue Z, Nie L, Zhao P, Ji N, Liao G and Wang Q (2022) Senescence-associated secretory phenotype and its impact on oral immune homeostasis. Front. Immunol. 13:1019313. https://doi.org/10.3389/fimmu.2022.1019313.

Chapter 2
Genomic Instability

Priscila Farias Tempaku

Maintenance of Genomic Integrity

Throughout evolution, the genome has faced a constant risk of harm from both internal and external sources. These threats include oxidative species generated by cellular metabolic processes, spontaneous alterations or losses of genetic bases, exposure to radiation, and various chemicals. Consequently, over time, mechanisms to manage and repair DNA damage have emerged and remained remarkably conserved. Organisms require a degree of genome stability to maintain their essential functions, but they must also permit controlled mutational changes to facilitate adaptation and drive the process of evolution [1].

The main guardians of genome integrity encompass three main mechanisms: metabolizing enzymes, DNA repair pathways and signaling molecules [2].

As the initial line of defense, key antioxidant defenses include glutathione, enzymes like glutathione peroxidase, superoxide dismutase, and catalase. Also, phase I and phase II metabolizing enzymes play a critical role in eliminating toxic substances from the body.

The second line of defense, DNA repair mechanisms, addresses DNA damage that persists despite other safeguards. Repair processes include:

- Nucleotide Excision Repair (NER): This repair mechanism addresses distortions in the DNA double helix and damage to the single strand. NER removes damaged nucleotides in blocks and replaces them with new ones, precisely at the site of the DNA damage.
- Base Excision Repair (BER): It comes into play when mutagenic agents impact the nitrogenous bases. BER eliminates the faulty base through a cleavage process

P. F. Tempaku (✉)
Universidade Federal de São Paulo, Rua Napoleão de Barros, 925, Vila Clementino, São Paulo/SP 04024-002, Brazil
e-mail: p.tempaku@gmail.com

V. Bueno (ed.), *Cellular and Molecular Aspects of Ageing*,
https://doi.org/10.1007/978-3-031-55022-5_2

of the nitrogenous base (deoxyribose), followed by the insertion of the correct base by the action of DNA polymerase.
- Mismatch Repair (MMR): This repair system is responsible for rectifying errors that occur during DNA replication and gene recombination, resulting in the mispairing of nucleotides.

In regard to DNA damage signaling, reactive oxygen species (ROS) not only cause direct DNA damage but also activate signal transduction pathways and transcription factors, including p53 and Nrf2. Nrf2 regulates defensive and detoxification genes, playing a dual role in cancer development [3]. It acts as both an anticancer protein, upregulated by tumor suppressor proteins, and a tumor promoter, overexpressed in many cancers and linked to resistance to chemotherapeutic drugs [4]. NF-κB is an inducible transcription factor involved in inflammation and cancer promotion. Deregulation of NF-κB is associated with cancer development, as it controls the expression of numerous pro-inflammatory, anti-apoptotic, and metabolism-related genes.

Causes of Genomic Instability

The occurrence of genomic instability can be attributed to various factors, either inherited or acquired, that affect essential processes such as DNA repair, DNA replication, cell cycle control, and chromosome segregation. Genomic instability becomes evident through an increased frequency of mutations in the genome, including alterations in the nucleotide sequence (such as base substitutions, base loss, nucleotide deletion, insertion, or amplification), DNA breaks, chromosomal translocations, inversions, deletions, and irregular chromosome numbers (known as aneuploidy) [5].

External sources of DNA damage, stemming from exposure to environmental chemicals, nanoparticles (NPs), radiation, or biological agents, can compromise genome integrity. Additionally, endogenously induced DNA damage, often as a result of metabolic processes, is a notable contributor. It's estimated that every human cell's genome accumulates thousands of DNA lesions of endogenous origin on a daily basis.

Furthermore, alterations in the epigenome can indirectly induce genome instability. Epigenetic changes have the potential to impact DNA repair efficiency and fidelity by influencing the expression of DNA repair genes. They may also induce disruptions in chromatin structure and deregulate enzymes involved in antioxidant defenses or xenobiotic metabolism, further contributing to genome instability.

Types of Genomic Instability

The DNA lesions can lead to the formation of single-strand breaks (SSBs) or double-strand breaks (DSBs). DNA repair is of utmost importance in maintaining organismal integrity and ensuring genomic stability. SSBs can be repaired through the base excision repair or nucleotide excision repair pathways, where the damage is removed, and the original DNA sequence is restored by the DNA polymerase enzyme, using the undamaged strand as a template [6].

On the other hand, DSBs are cytotoxic lesions that, if left unrepaired, could lead to genomic instability, cancer, and even cell death. They are more perilous since both strands of the DNA double helix are broken, leaving no intact template strand for repair. Consequently, the repair pathways are not considered error-free, and erroneous DSB repair can result in chromosomal rearrangements and loss of heterozygosity [7].

Although DSB repair is critical for genome integrity maintenance, the repair process requires close and precise regulation and monitoring. The two most used pathways for repairing DSBs in higher eukaryotes include non-homologous end-joining (NHEJ) and homologous recombination (HR). NHEJ is prone to errors and serves as a rapid means to seal the broken DNA ends and restart replication. In contrast, HR is a high-fidelity process.

Genomic instability can manifest in various forms, including:

- Point Mutations: Single-base changes in the DNA sequence, such as substitutions, insertions, or deletions.
- Chromosomal Rearrangements: Alterations in the structure or arrangement of chromosomes, which may include translocations, inversions, duplications, or deletions.
- Copy Number Variations (CNVs): Changes in the number of copies of specific DNA segments, leading to variations in gene dosage.
- Microsatellite Instability: Occurs when short, repetitive DNA sequences within the genome exhibit changes in length due to replication errors or deficiencies in DNA mismatch repair.

Consequences of Genomic Instability

In the event of a repair failure and the accumulation of damage, three potential scenarios may unfold: cells have the potential to undergo transformation and develop into cancerous forms, cells can transition into a non-proliferating state known as senescence, or cells may undergo programmed cell death, as seen in apoptosis, for instance.

Programmed cell death in the form of apoptosis is a vital process for regular cell turnover and is integral to various biological functions. Apoptosis can occur through the activation of caspases, signals from death receptors on the cell membrane, or induction by granzyme B released by cytotoxic T cells (Tc cells). Active caspases trigger endonucleases and proteases, ultimately resulting in cell death [8].

When a cell manages to evade death in the face of DNA damage, it may enter a state of senescence instead, lingering in an unhealthy and proinflammatory condition. Cells can enter senescence due to various triggers, including replicative senescence in somatic cells, oncogene-induced senescence, and excessive DNA damage [9].

Clinical Consequences

Genomic instability has profound implications for an organism's health, development, and evolution. Several conditions have been associated with the mechanisms of genomic instability such as cancer, ageing and age-related diseases, neurodegenerative disorders and autoimmune conditions [10–12].

One of the most studied conditions associated with genomic instability is cancer, where accumulated mutations can lead to uncontrolled cell growth and the formation of malignant tumors. The process of tumorigenesis is closely related to the phenomenon of genomic instability. Transformed cells may exhibit chromosomal rearrangements, as well as the loss or gain of entire chromosomes, leading to chromosomal instability and contributing to tumor progression. Replicative stress, which is the condition where the replication fork progresses slowly, can be one of the causes of genomic instability. The overexpression of oncogenes, loss of tumor suppressor genes, or dysregulation of genes involved in DNA damage signaling pathways can lead to a condition of replicative stress, which can interfere with genome integrity and contribute to the tumorigenesis process [13].

Evidence in the literature also supports that neurodegenerative diseases are closely related to genomic instability. Currently, there are several theories that attempt to elucidate the molecular underpinnings of neurodegeneration. Among these, a particular focus has been placed on somatic mutations and genomic instability in brain cells. Numerous studies have also suggested that the pathological mechanisms observed in neurodegenerative diseases can give rise to genomic instability, including both genomic and chromosomal instability. These processes encompass issues such as compromised DNA repair, the inappropriate activation of neuronal cell cycles, replicative stress, and disruptions in the regulation of chromosome segregation during mitosis. Notably, Alzheimer's disease, one of the most prevalent late-onset neurodegenerative diseases, shares similarities with malignancy in terms of its underlying processes. These processes can lead to either neuronal cell death or a state resembling cellular senescence. It's worth mentioning that the prevalence of aneuploidy in brain cells, specifically the loss of the X chromosome, a cytogenetic marker of aging, increases by more than two-fold in Alzheimer's disease [10].

As the focus of this chapter, the next topic will review with further detail the association between genomic instability and ageing.

The Association Between Genomic Instability and Ageing

Ageing is a universal biological phenomenon characterized by a gradual and irreversible deterioration in the physical function of all organ systems. This deterioration is primarily driven by the accrual of damage in response to various stressors. At the cellular level, ageing is associated with a range of molecular and biochemical changes [14].

In 2013, López-Otín and colleagues provided a comprehensive overview of aging in various organisms, particularly mammals, outlining nine common features. In 2023, these features were updated and currently encompass twelve factors: genomic instability, telomere shortening, epigenetic alterations, loss of proteostasis, disabled macroautophagy, deregulated nutrient sensing, mitochondrial dysfunction, cellular senescence, stem cell exhaustion, altered intercellular communication, chronic inflammation and dysbiosis. All of these attributes are highly intertwined; thus, the modulation of one affects others [15].

As mentioned previously, genomic instability arises from an imbalance between DNA damage and repair mechanisms. During the ageing process there is a diminished intrinsic efficiency of DNA repair pathways, a process that is delayed by dietary restriction [16].

Evidence in the literature have pointed out a decline in DNA damage response (DDR) capacity during the ageing process. Markers of DNA damage have been observed in age-associated diseases such as dementias, cardiovascular disease, and cancer, suggesting that genome instability could be a causal factor in these pathologies. One example of this decline is the repair of double-strand breaks (DSBs). The repair process involves two major pathways: Homologous Recombination and Non-Homologous End Joining (NHEJ). In general, the absence of repair in this context has been associated with an increased occurrence of genomic rearrangements due to the accumulation of mutations [17].

A reduced ability to repair DSBs has been observed in the peripheral lymphocytes of elderly patients, as well as a decrease in repair levels in ageing tissues and senescent cells [18]. It is worth mentioning that the decline in DDR capacity varies among different tissues. For instance, NHEJ activity was examined in various tissue samples from young and aged mice, revealing an overall decrease in activity. Concurrently, an increase in alternative NHEJ (alt-NHEJ), a version of NHEJ more prone to induce mutagenic effects, became more prominent. Similar effects have recently been reported in yeast models during replicative ageing, as well as in Drosophila and mice, underscoring the conserved nature of DSB and other DNA repair deficiencies in aging cells. Additionally, the capacity for nucleotide excision repair (NER) is known to diminish during nematode ageing.

The connection between ageing and a decline in the DNA damage response (DDR) capacity is further supported by progeria syndromes or premature ageing disorders, which replicate many characteristics associated with ageing in young individuals [19–22].

Inherited defects in DNA repair pathways are linked to specific genome instability syndromes characterized by developmental abnormalities, a heightened risk of cancer, and signs of accelerated ageing. Hutchinson–Gilford, Werner, Cockayne, and XFE progeroid syndrome are examples of diseases causing severe accelerated ageing and are associated with genome instability. For example, mutations in the A-type nuclear lamina LMNA result in genome stability impairment in Hutchinson–Gilford progeroid syndrome. Null mutations in the WRN gene, which encodes a RECQ DNA helicase with critical roles in DNA repair and recombination, cause accelerated genome instability in Werner syndrome. Various genetic alterations collectively lead to a dysfunction of the nucleotide excision DNA repair system in Cockayne syndrome. XFE progeroid syndrome is a result of mutations in ERCC4/FANCQ, leading to reduced expression of XPF-ERCC1, a heterodimeric DNA repair endonuclease essential for NER and the repair of some DNA DSBs. These syndromes collectively illustrate the devastating impact of DNA damage on human health and serve as important experimental models for ageing research. Progeroid syndromes result from various mutations affecting the expression or activity of key proteins involved in DDR.

Interestingly, the significance of an efficient DNA repair system for achieving longevity has been underscored by a study conducted by Garagnani and colleagues on supercentenarians (110 years old) and semi-supercentenarians (105 years old). Through whole-genome sequencing and a comparison with young controls from the same geographic regions, they identified an increased activity of DNA repair genes in the older cohort, as well as lower mutational patterns compared to their younger counterparts. As a result, reinforcing DDR capacity has been shown to have positive effects on longevity in various models [23]. Comparing the ageing process across different rodent species, the sirtuin-regulated strength of DSB repair has been linked to longevity, while the overexpression of DDR genes in Drosophila has been demonstrated to enhance lifespan [24].

As cells age, the nature of apoptotic activity changes. Increased apoptosis has been observed in the heart, kidney, skeletal muscle, and Tc cells, potentially contributing to decreased cellularity in these tissues. This increase across various tissues may be associated with elevated production of free radicals and exacerbated by the accumulation of DNA damage in aging cells. As the risk of cells becoming cancerous and dysfunctional rises with age, increased apoptosis in aged cells is suggested to be a defensive mechanism. In contrast, in some tissues like the colon, apoptosis seems to decrease with age, potentially leading to the accumulation of senescent cells and age-related cancer development.

Another form of cell death known as parthanatos is intriguing because it is closely linked to DNA damage. In this case, activation of the DNA damage responder PARP1 serves as the initial trigger, leading to the formation of PAR polymers, activation of apoptosis-inducing factor (AIF), and caspase-independent cell death. Parthanatos has been implicated in age-related neurodegeneration, particularly in Parkinson's disease, a condition where DNA damage has been observed. Accordingly, markers of double-strand DNA breaks have been detected in the neurons of Alzheimer's and Parkinson's patients, suggesting that age-related DNA damage might contribute to neurodegeneration through parthanatos. Parthanatos has also been associated with chronic heart failure and diabetes [25].

At younger ages, senescence plays a role in tumor suppression, wound healing, and tissue development. However, as one ages, these senescent cells accumulate and are likely to contribute to various health issues. Senescent cells adopt a proinflammatory secretory profile known as the senescence-associated secretory phenotype, increasing the secretion of inflammatory cytokines and contributing to a proinflammatory microenvironment [26]. Overall, p53, p21, and p16 are key players in initiating senescence in response to excessive DNA damage. It seems that signaling from the site of DNA damage, rather than the damage itself, is the trigger for entry into senescence once a certain level of genotoxic stress is reached. Consequently, point mutations, as seen in mismatch repair deficiencies that do not activate the canonical DNA damage response, are not associated with senescence. In contrast, DNA breaks or helix-distorting lesions that activate the DNA damage response can induce senescence. These findings suggest that the DNA damage response may drive degenerative processes like neurodegeneration and sarcopenia, while point mutations may go unnoticed without affecting cell growth. Hyperactivation of the DNA damage response is linked to neurodegeneration, whereas hypermutability, as observed in the mismatch DNA repair-defective disease Lynch syndrome, is associated with cancer development. Supporting this hypothesis is the observation that mutations in p53 or p16, which facilitate senescence signaling, are among the most frequently mutated genes in cancers [27].

Interestingly, when senescent cells are removed, either through genetic manipulation or with senolytic drugs, biological aging is significantly slowed in mice. Consequently, there are ongoing trials to assess the potential of senolytics in postponing age-associated health issues in humans. It's noteworthy that multiple drugs are being explored that either directly or indirectly affect DNA repair or the consequences of DNA damage [28].

Perspectives

Fortunately, various DNA repair pathways are in place to rectify the lesions associated with genome instability, preserving the integrity of the genome. Gene mutations, characterized by irreversible changes in the nucleotide sequence, are relatively rare and typically arise due to misrepair or erroneous replication of DNA sequences containing lesions. When mutations affect genes crucial for maintaining precise DNA integrity, it can lead to altered gene expression, including oncogenes or tumor suppressor genes, with genome instability being one of the resulting consequences.

Genomic instability can have far-reaching consequences, particularly in the context of human health. It is a driving force behind the development of various diseases. Accumulated mutations and genomic alterations can lead to uncontrolled cell growth and the formation of malignant tumors. Additionally, genomic instability is associated with ageing and age-related diseases, neurodegenerative disorders, and autoimmune conditions.

In the context of ageing, evidence points out that genome instability is a crucial factor in the progression of ageing, and safeguarding our ageing genomes is of utmost importance for promoting healthy ageing. Diseases characterized by premature ageing could offer a promising venue for testing ageing interventions and assessing outcomes more efficiently and cost-effectively. In this direction, a multitude of therapies are currently emerging, either directly or indirectly contributing to reduced DNA damage. The extensive ongoing global research in this area will undoubtedly pave the way for interventions that benefit longevity.

References

1. Kovalchuk I. Chapter 1 - Genome Stability: An Evolutionary Perspective, Academic Press, 2016, Pages 1–18, ISBN 9780128033098. doi: https://doi.org/10.1016/B978-0-12-803309-8.00001-X.
2. Wang J, Lindahl T. Maintenance of Genome Stability. Genomics Proteomics Bioinformatics. 2016 Jun;14(3):119–121. doi: https://doi.org/10.1016/j.gpb.2016.06.001. Epub 2016 Jun 18. PMID: 27329679; PMCID: PMC4936602.
3. Bhattacharjee S, Nandi S. Choices have consequences: the nexus between DNA repair pathways and genomic instability in cancer. Clin Transl Med. 2016 Dec;5(1):45. doi: https://doi.org/10.1186/s40169-016-0128-z. Epub 2016 Dec 5. PMID: 27921283; PMCID: PMC5136664.
4. Bartkova J, Horejsí Z, Koed K, et al. DNA damage response as a candidate anti-cancer barrier in early human tumorigenesis. Nature. 2005 Apr 14;434(7035):864–70. doi: https://doi.org/10.1038/nature03482. PMID: 15829956.
5. Langie SA, Koppen G, Desaulniers D, et al. Causes of genome instability: the effect of low dose chemical exposures in modern society. Carcinogenesis. 2015 Jun;36 Suppl 1(Suppl 1):S61–88. doi: https://doi.org/10.1093/carcin/bgv031. PMID: 26106144; PMCID: PMC4565613.
6. Caldecott KW. Single-strand break repair and genetic disease. Nat Rev Genet. 2008 Aug;9(8):619–31. doi: https://doi.org/10.1038/nrg2380. PMID: 18626472.
7. White RR, Vijg J. Do DNA Double-Strand Breaks Drive Aging? Mol Cell. 2016 Sep 1;63(5):729–38. doi: https://doi.org/10.1016/j.molcel.2016.08.004. PMID: 27588601; PMCID: PMC5012315.
8. Elmore S. Apoptosis: a review of programmed cell death. Toxicol Pathol. 2007 Jun;35(4):495–516. doi: https://doi.org/10.1080/01926230701320337. PMID: 17562483; PMCID: PMC2117903.
9. Jeyapalan JC, Ferreira M, Sedivy JM, Herbig U. Accumulation of senescent cells in mitotic tissue of aging primates. Mech Ageing Dev. 2007 Jan;128(1):36–44. doi: https://doi.org/10.1016/j.mad.2006.11.008. Epub 2006 Nov 20. PMID: 17116315; PMCID: PMC3654105.
10. Iourov IY, Vorsanova SG, Kurinnaia OS, Zelenova MA, Vasin KS, Yurov YB. [Causes and Consequences of Genome Instability in Psychiatric and Neurodegenerative Diseases]. Mol Biol (Mosk). 2021 Jan-Feb;55(1):42–53. Russian. doi: https://doi.org/10.31857/S0026898421010158. PMID: 33566024.
11. Vijg J, Suh Y. Genome instability and aging. Annu Rev Physiol. 2013;75:645–68. doi: https://doi.org/10.1146/annurev-physiol-030212-183715. PMID: 23398157.
12. De Majo F, Martens L, Hegenbarth JC, et al. Genomic instability in the naturally and prematurely aged myocardium. Proc Natl Acad Sci U S A. 2021 Sep 7;118(36):e2022974118. doi: https://doi.org/10.1073/pnas.2022974118. PMID: 34465617; PMCID: PMC8433522.

13. Gorgoulis VG, Vassiliou LV, Karakaidos P, et al. Activation of the DNA damage checkpoint and genomic instability in human precancerous lesions. Nature. 2005 Apr 14;434(7035):907–13. doi: https://doi.org/10.1038/nature03485. PMID: 15829965.
14. Guo J, Huang X, Dou L, et al. Aging and aging-related diseases: from molecular mechanisms to interventions and treatments. Signal Transduct Target Ther. 2022 Dec 16;7(1):391. doi: https://doi.org/10.1038/s41392-022-01251-0. PMID: 36522308; PMCID: PMC9755275.
15. López-Otín C, Blasco MA, Partridge L, et al. Hallmarks of aging: An expanding universe. Cell. 2023 Jan 19;186(2):243–278. doi: https://doi.org/10.1016/j.cell.2022.11.001. Epub 2023 Jan 3. PMID: 36599349.
16. Dempsey JL, Pfeiffer M, Morley AA. Effect of dietary restriction on in vivo somatic mutation in mice. Mutat Res. 1993 Apr;291(2):141–5. doi: https://doi.org/10.1016/0165-1161(93)90153-q. PMID: 7680418.
17. Tucker JD, Spruill MD, Ramsey MJ, et al. Frequency of spontaneous chromosome aberrations in mice: effects of age. Mutat Res. 1999 Mar 10;425(1):135–41. doi: https://doi.org/10.1016/s0027-5107(99)00036-6. PMID: 10082924.
18. Jones IM, Thomas CB, Tucker B, et al. Impact of age and environment on somatic mutation at the hprt gene of T lymphocytes in humans. Mutat Res. 1995 Oct;338(1–6):129–39. doi: https://doi.org/10.1016/0921-8734(95)00018-2. PMID: 7565868.
19. Dollé ME, Giese H, Hopkins CL, et al. Rapid accumulation of genome rearrangements in liver but not in brain of old mice. Nat Genet. 1997 Dec;17(4):431–4. doi: https://doi.org/10.1038/ng1297-431. PMID: 9398844.
20. Dollé ME, Snyder WK, Gossen JA, et al. Distinct spectra of somatic mutations accumulated with age in mouse heart and small intestine. Proc Natl Acad Sci U S A. 2000 Jul 18;97(15):8403–8. doi: https://doi.org/10.1073/pnas.97.15.8403. PMID: 10900004; PMCID: PMC26960.
21. Lee JW, Ong EBB. Genomic Instability and Cellular Senescence: Lessons From the Budding Yeast. Front Cell Dev Biol. 2021 Jan 12;8:619126. doi: https://doi.org/10.3389/fcell.2020.619126. PMID: 33511130; PMCID: PMC7835410.
22. Garcia AM, Busuttil RA, Calder RB, et al. Effect of Ames dwarfism and caloric restriction on spontaneous DNA mutation frequency in different mouse tissues. Mech Ageing Dev. 2008 Sep;129(9):528–33. doi: https://doi.org/10.1016/j.mad.2008.04.013. Epub 2008 May 13. PMID: 18565572; PMCID: PMC2581895.
23. Garagnani P, Marquis J, Delledonne M, et al. Whole-genome sequencing analysis of semi-supercentenarians. Elife. 2021 May 4;10:e57849. doi: https://doi.org/10.7554/eLife.57849. PMID: 33941312; PMCID: PMC8096429.
24. Shaposhnikov M, Proshkina E, Shilova L, Zhavoronkov A, Moskalev A. Lifespan and Stress Resistance in Drosophila with Overexpressed DNA Repair Genes. Sci Rep. 2015 Oct 19;5:15299. doi: https://doi.org/10.1038/srep15299. PMID: 26477511; PMCID: PMC4609912.
25. Wang Y, Luo W, Wang Y. PARP-1 and its associated nucleases in DNA damage response. DNA Repair (Amst). 2019 Sep;81:102651. doi: https://doi.org/10.1016/j.dnarep.2019.102651. Epub 2019 Jul 8. PMID: 31302005; PMCID: PMC6764844.
26. Coppé JP, Desprez PY, Krtolica A, Campisi J. The senescence-associated secretory phenotype: the dark side of tumor suppression. Annu Rev Pathol. 2010;5:99–118. doi: https://doi.org/10.1146/annurev-pathol-121808-102144. PMID: 20078217; PMCID: PMC4166495.
27. Petr MA, Tulika T, Carmona-Marin LM, Scheibye-Knudsen M. Protecting the Aging Genome. Trends Cell Biol. 2020 Feb;30(2):117–132. doi: https://doi.org/10.1016/j.tcb.2019.12.001. Epub 2020 Jan 6. PMID: 31917080.
28. Friedberg EC. Suffering in silence: the tolerance of DNA damage. Nat Rev Mol Cell Biol. 2005 Dec;6(12):943–53. doi: https://doi.org/10.1038/nrm1781. PMID: 16341080.

Chapter 3
Telomeres

Flavia S. Donaires and Rodrigo T. Calado

Telomeres at a Glance

Telomeres, repetitive DNA sequences and associated protein complexes at the ends of eukaryotic chromosomes are critical guardians of genome integrity and play an essential role in cellular longevity. These specialized structures undergo dynamic changes during cellular divisions and act as protective caps, preventing the loss of coding genetic material (due to peculiarities of the replication process) and the fusion of chromosomal ends. They also participate in cellular processes such as DNA replication, repair, and senescence. Telomere length maintenance is tightly regulated by telomerase and alternative lengthening of telomeres (ALT) mechanisms, which have been extensively investigated in physiological aging and diseases. In humans, the antagonistic phenomena of excessive telomere erosion or elongation may translate into degenerative diseases (telomeropathies) and cancer. This chapter provides a comprehensive overview of telomere biology, including their structural composition, functions, maintenance, and the intricate interplay between telomeres and cellular processes crucial for preserving genetic information.

A Long History for a Short Repetitive Sequence

The cells of eukaryotic organisms have a nucleus, a membrane-bound compartment distinct from other cellular organelles that contain the genetic material in deoxyribonucleic acid (DNA). The DNA molecules are wrapped around proteins, the histones, forming long, thread-like structures termed chromosomes. The ends of

F. S. Donaires · R. T. Calado (✉)
Department of Medical Imaging, Hematology, and Oncology, Ribeirão Preto Medical School, University of São Paulo, Ribeirão Preto, Brazil
e-mail: rtcalado@fmrp.usp.br

V. Bueno (ed.), *Cellular and Molecular Aspects of Ageing*,
https://doi.org/10.1007/978-3-031-55022-5_3

chromosomes are called *telomeres* (from the Greek terms *telos* and *meros*, meaning "end" and "part", respectively).

The term "telomere" was introduced in 1938 by Hermann Muller, who observed that *Drosophila* chromosomes with radiation-induced breaks were not transmitted to progeny [40]. A couple of years later, Barbara McClintock's experiments revealed that the newly generated ends of broken maize chromosomes exhibited adhesive properties and a tendency to fuse. Conversely, the natural ends of chromosomes remained stable and did not fuse with other broken or intact ends [35]. These findings suggested that telomeres possess specialized structures that distinguish them from breakage-induced chromosome ends, inhibiting genetic recombination. However, neither Muller nor McClintock had sufficient tools to elucidate the molecular details of the telomere structure. This structure was understood in the 1980s, firstly in the ciliate *Tetrahymena,* through the collaborative efforts of researchers such as Elizabeth Blackburn, Carol Greider, and Jack Szostak. Blackburn and Szostak were the pioneers in hypothesizing the existence of a repetitive DNA sequence at the ends of chromosomes (revised in [5]).

Although slightly different among species, the telomeric DNA sequences have a basic G-rich repeat unit in the leading strand with the pattern $5' \ T_{1\text{–}4}A_{0\text{–}1}G_{1\text{–}8}$ $3'$. The pattern in humans and other vertebrates is the highly conserved hexameric TTAGGG sequence [39]. The extension of the repetitive sequences varies among different species. In humans, telomeres typically range from 5 to 15 kilobases (kb), with variations among individuals and within the same individual, depending on the cell type and age.

The very end of telomeres has a protrusion with a free hydroxyl in the G-rich leading strand (the 3′ overhang) formed due to the catalytic addition of telomeric repeats and the post-replicative processing of the lagging strand [31]. This protrusion (containing 50–500 bases in humans) invades and displaces a segment of the double-stranded telomeric DNA, forming a looped structure known as the D loop. In addition, the single-stranded overhang folds back in the double-stranded repeats, base pairing with part of the C-rich strand, forming a structure called T loop [17].

The telomeric repeats are capped with a protein complex, the shelterin, resulting in a tertiary structure [17]. This complex is composed of six proteins: Telomeric Repeat Binding Factor 1 and 2 (TRF1 and TRF2, respectively), Protection of Telomeres 1 (POT1), TRF1-interacting Nuclear Factor 2 (TIN2), Tripeptidyl Peptidase 1 (TPP1 or ACD), and TRF2-interacting Protein (RAP1). Three directly bind to the telomeric nucleotides, TRF1 and TRF2 to double-stranded repeats and POT1 to single-stranded repeats. TIN2 connects TRF1 and TRF2 to the rest of the shelterin components. TPP1 forms a complex with POT1 and enhances its affinity for the single-stranded telomeric DNA. While RAP1 doesn't directly interact with DNA, it binds to TRF2 and contributes to the formation of the shelterin complex. Shelterin proteins mediate critical functions in telomere maintenance, such as telomere length regulation as well as formation and integrity of the T loops [19, 51].

Together, the telomeric repeats and the shelterin provide structural features and protein interactions that protect the end of the linear DNA molecule to ensure chromosome stability through at least three unique properties, discussed below.

First, telomeres prevent deoxyribonucleases (DNases) from degrading the ends of the linear DNA molecules. Telomeres are maintained in a heterochromatic state, a tightly packed chromatin structure less accessible to DNA-degrading enzymes. In addition, the single-stranded overhang is hidden from DNases and other DNA repair machinery proteins through the T loop formation and the POT1 binding. Other shelterin, such as TRF2, prevents DNA repair proteins from recognizing telomeres as double-strand breaks. TIN2 indirectly prevents aberrant processing and degradation of telomeres by coordinating the interactions of shelterin, ensuring the proper spatial arrangement of the telomere components [17].

Second, telomeres prevent the fusion of the ends with other DNA molecules. Again, T loops and shelterin play a role in this feature, mediating the suppression of DNA damage response and repair pathways. T-loop configuration conceals the end of the telomere and prevents it from being recognized as a DNA break, thus minimizing the chances of recombination or rearrangements. POT1 binding prevents the overhang from interacting with other single-stranded DNA regions, reducing the probability of fusion events. TRF1 and TRF2 interact with one another and with telomeric DNA, keeping telomeres separate [17].

Third, telomeres facilitate replication of the ends of the linear DNA without losing genetic information due to the peculiarities of the replication mechanism. DNA polymerase requires an RNA primer with a 3′-hydroxyl donor to replicate the lagging strand. As primer binding becomes impossible at the very end portion of the strand, a gap is left at the chromosomal termini. Without a compensatory mechanism, the linear DNA molecule would gradually lose several bases during each mitotic division [15, 21]. This biological challenge, the "end-replication problem", was sighted by Alexey Olovnikov [44]. Telomeres solve the coding genetic sequence loss issue during cell division; however, in highly proliferative tissues, all telomeric sequences would ultimately be depleted, raising the problem of linear chromosome ends with no protective mechanism. Indeed, in cells lacking the activity of mechanisms for telomere length maintenance, telomeres gradually shorten after each round of replication, with a loss estimated in 50–150 base pairs of telomeric DNA in human fibroblasts, for instance [34]. In human peripheral blood leukocytes, telomere attrition is mathematically modeled as a cubic polynomial function, as it is more abrupt in the first and last decades of life but attenuated between the twentieth and fiftieth decades (Fig. 3.1, upper graph). Sex hormones influence this attenuation [10].

Upon reaching a critically short length, telomeres trigger cellular signaling pathways leading to proliferation arrest, apoptosis, or senescence [25]. That explains Leonard Hayflick's conclusion about the limited proliferative capacity of cells when he was studying cells in culture for several passages [22]. His observation was postulated as the "Hayflick limit" and mechanistically associated with telomere erosion [49]. However, germline cells, embryonic and adult stem cells, as well as immortalized and cancer cells, surpass the expected Hayflick limit, exhibiting a higher proliferation capacity as they can maintain their telomere length to some degree.

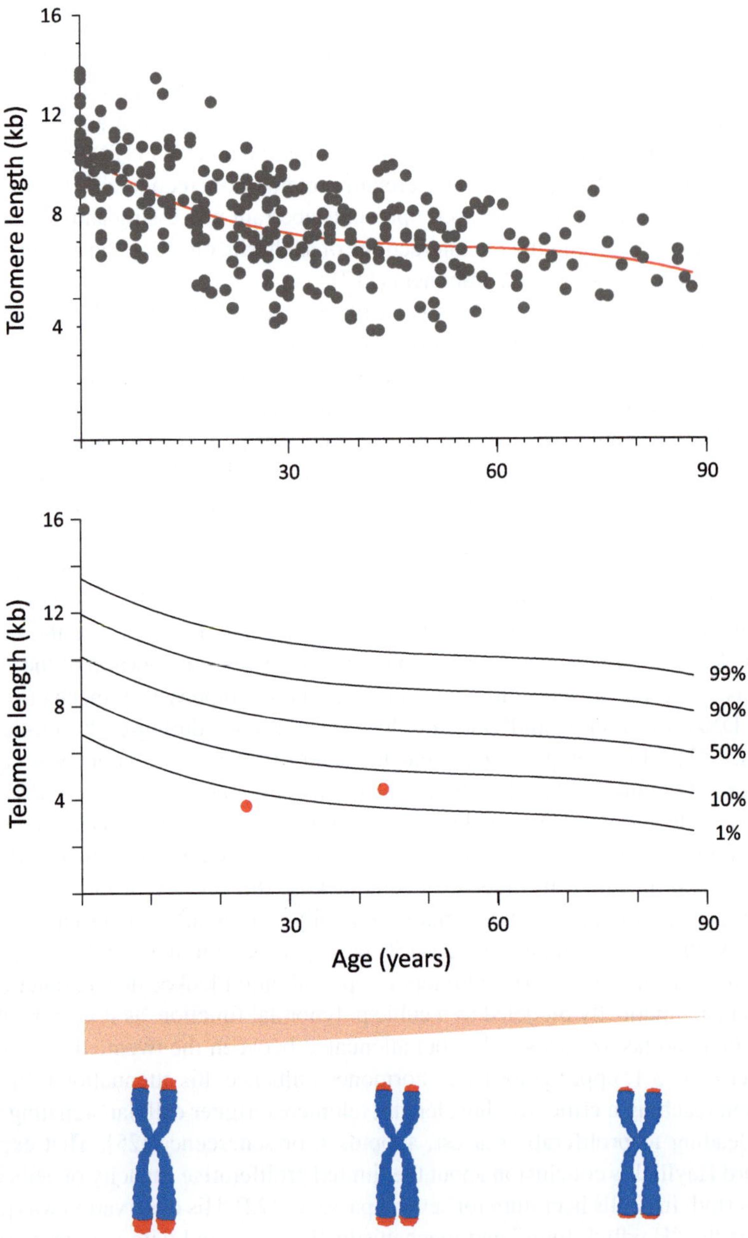

Fig. 3.1 Telomere length in human leukocytes as a function of age. Telomere shortening occurs naturally with aging, and is mathematically represented as a cubic polynomial function (upper graph). The expected telomere length for an individual within a certain age is calculated using curves derived from telomere lengths of several healthy individuals from a population; age-adjusted telomeres are considered short or very short if below the tenth or the first percentile, respectively (lower graph). Telomeres are represented in kilobases (kb)

Here Is a Tip: How to Maintain Telomeres to Go Longer

Intrigued with the observations that telomeres had a replication mechanism as the repetitive sequences were somehow added at the chromosomal termini of the model organisms *Tetrahymena*, yeast, and trypanosomes, Blackburn's lab started biochemical experiments looking for evidence for a telomere-addition enzyme. In 1985, Blackburn's graduate student Carol Greider identified the telomerase RNA component (TERC) in *Tetrahymena*, while Szostak's work revealed the role of *EST2* in yeast telomere maintenance. Carol Greider's identification of a TERC-EST2 complementary sequence and subsequent purification of the protein from *Tetrahymena* led to the in vitro demonstration of a protein, termed telomerase, with the ability to extend telomeres. By the early 1990s, Greider's team established human telomerase activity (revised in [5]). In recognition of their achievements, Blackburn, Greider, and Szostak were awarded the Nobel Prize in Physiology or Medicine in 2009.

The human telomerase is a ribonucleoprotein that consists of two main components: the telomerase reverse transcriptase (TERT), which is a catalytic protein subunit, and TERC, which provides the template for the synthesis of the telomeric DNA. The repetitive DNA sequences are de novo synthesized in the 3′ end of the single-stranded overhang through the reverse transcription of the RNA template by TERT.

In vertebrates, the TERC's 3′ portion contains an H/ACA domain with a CAB box motif, which characterizes this molecule as a small Cajal body-specific RNA (scaRNA). The proteins dyskerin (encoded by *DKC1*), NOP10, NHP2, and GAR1 bind to the H/ACA site to stabilize TERC. TCAB1 (encoded by *WRAP53*) binds to dyskerin and the TERC's CAB site, driving the complex to subnuclear structures called Cajal bodies, where TERC is maturated, and the telomerase components are assembled before transported to telomeres [54]. The stoichiometry of the functional holoenzyme is a single subunit of TERT, a subunit of TERC assembled into two sets of H/ACA core proteins, along with a single subunit of TCAB1 [18].

Telomerase activity is tightly regulated: it is highly active during embryonic development, low or absent in most adult somatic cells, and often reactivated in certain cancer cells, contributing to their immortality and uncontrolled growth. This regulation is a complex process involving multiple layers of control. At the transcriptional level, the expression of telomerase components, especially *TERT*, is proportional to the necessity of telomerase activity. *TERC* is the most sensitive component at the post-transcriptional level, as its maturation, stability, and transport depend on several other structures. An essential protein in these processes is PARN, an exonuclease with deadenylation activity required for TERC maturation, which plays a role in maintaining its half-life and steady-state levels [38].

Protein–protein interactions between telomerase components and shelterin also influence telomerase activity. TPP1, POT1, and TRF1/TRF2 interact with telomerase to facilitate or inhibit its access to telomeric DNA. Similarly, changes in chromatin structure and epigenetic modulation, such as histone modifications and DNA methylation at the telomeres, affect telomerase access to telomeric DNA.

Another modulator of telomerase activity is its localization. The holoenzyme is recruited to telomeres during the S-phase when DNA replication occurs. Cell-cycle checkpoints and DNA damage response (DDR) pathways can inhibit telomerase, preventing cells with damaged DNA from replicating uncontrollably. DDR pathways, particularly the Ataxia Telangiectasia-Mutated (ATM) and Ataxia Telangiectasia and Rad3-related (ATR) pathways, also become activated in response to critically short telomeres. These pathways may induce cell cycle arrest or inhibit telomerase via ATM or p53 activation. ATM kinase can phosphorylate and modify telomerase-associated proteins, hindering telomerase to access and extend telomeres. Activated p53, regulated by ATM and ATR, can suppress telomerase by inhibiting *TERT* transcription [41].

On the other hand, telomerase activity may be suppressed when telomeres are long enough and structurally intact. In this scenario, the shelterin complex, especially TRF2, physically blocks the telomerase enzyme from interacting with the telomeric DNA, through direct interaction or indirectly, as it assists T loop formation. If telomeres become aberrantly long and shelterin becomes sparse along the strand, there is a mechanism of trimming stimulated by the Telomeric Zinc Finger-Associated Protein (TZAP). This protein competes with the shelterin complex to telomeric DNA binding and triggers T loop excision [30].

Other proteins, such as RTEL1, a DNA helicase that disassembles telomeric structures (D and T loops), also participate in telomere maintenance. RTEL1 disrupts the D loop formation during homologous recombination [29, 52, 56] and unwinds the T loops during DNA replication to make telomeres accessible for the telomerase [53].

Telomeric and subtelomeric regions were thought to be transcriptionally inactive for decades due to their primarily heterochromatic composition. However, Azzalin *et al.* demonstrated that the subtelomeric regions of several mammalian chromosomes are transcribed into long non-coding RNAs containing telomeric repeats that were defined as TERRA, an acronym for TElomeric-Repeat-containing RNA [2]. Thus, these transcripts comprise chromosome-specific sequences in the 5′ end, followed by 5′-UUAGGG-3′ repeats [42]. TERRA molecules participate in telomere maintenance through different mechanisms that, although still not completely elucidated, have been explored over the past years. Due to its portion that is complementary to the telomeres, TERRA can bind to TERC to either inhibit the activity of the enzyme [47] or form clusters that are recruited to short telomeres and promote telomere elongation [16].

The telomere length maintenance is mainly orchestrated by telomerase; however, 10–15% of cancer cells maintain telomeres via telomerase-independent alternative lengthening (ALT) mechanisms that are based on homologous recombination [13]. Besides cancer, telomerase and recombination pathways are balanced during embryonic development to guarantee telomere maintenance before the embryo starts differentiation [32, 46].

The Good, the Short, and the Long

Although telomere shortening is linked to aging, it is still unclear whether this characteristic is just a sign or a contributing factor. In humans, individuals with shorter telomeres live an average of five years less than those with longer telomeres. People older than 60 with shorter age-adjusted telomeres have a three-fold increase in heart disease and an eight-fold increase in infectious diseases [12]. Conversely, humans have significantly shorter telomeres than species like mice, which live only a few years—evidence that suggests telomeres alone do not determine lifespan.

Aging is a natural and multistep process in which DNA damage plays a central role [48]. Following this line of evidence, telomeres indirectly act in the aging process as a DNA integrity and stability protector and directly act as a mitotic clock, limiting the number of times a cell can divide. The importance of telomere homeostasis, which is the balance between telomere shortening and elongation, is evident when disrupted.

Telomere erosion compromises cell and, consequently, tissue renewal. The first connection between defects in telomere maintenance machinery and human disease was established in 1999 when short telomeres were found to be a defining characteristic of dyskeratosis congenita (DC). DC is a rare and progressive disorder that was primarily characterized as a triad of mucocutaneous manifestations: leukoplakia, abnormal skin pigmentation, and nail dystrophy. The first report dates from 1906 [57], but the molecular basis of the disease started with classical linkage analysis of large pedigrees in 1998 and the discovery that affected males harbored a defect in the X-linked gene *DKC1* [23]. A year later, Mitchell *et al.* [37] observed that dyskerin, until that time described with a putative nucleolar function, was associated not only with H/ACA small nucleolar RNAs but also with TERC, which contains a H/ACA RNA motif, finally associating decreased telomerase activity and short telomeres with the disease. The classical mucocutaneous phenotype of DC is accompanied by a high risk of bone marrow failure, pulmonary fibrosis, and liver disease, as well as a significantly increased risk of developing cancer. Although there is variability in clinical symptoms and organ involvement, bone marrow dysfunction is evident in over 80% of cases [27]. This dysfunction also represents a significant risk factor for mortality in these patients [9].

A broad DC-related phenotypic spectrum goes beyond the mucocutaneous triad, exhibiting variable symptoms and age of onset, even among individuals within the same family that share the genetic defect. Some patients present with complex multisystem disorders in early childhood (e.g., Coats plus, Hoyeraal-Hreidarsson, or Revesz syndromes), while others show a specific organ commitment in the youth or later in life (e.g., inherited aplastic anemia, familial pulmonary fibrosis, or liver cirrhosis), whereas others remain asymptomatic. The standard features in all these clinical presentations are age-adjusted short telomeres and/or a rare pathogenic germline variant in a telomere-biology gene. This group of diseases has been termed telomeropathies or telomere biology disorders [43]. The broad spectrum of phenotypes is attributed to variable genetic penetrance, expressivity, as well as environmental, epigenetic, and other genetic factors [9].

Over the past decades, advances in genomics techniques allowed the association of telomeropathies with pathogenic germline variants in at least sixteen telomere-biology genes (*ACD*, *CTC1*, *DKC1*, *NAF1*, *NHP2*, *NOP10*, *PARN*, *POT1*, *RPA1*, *RTEL1*, *STN1*, *TERC*, *TERT*, *TINF2*, *WRAP53*, and *ZCCHC8*), although about 20% of cases remain without a detectable genetic cause [43]. The most common mutated genes are *TERT*, *TERC*, *DKC1*, *RTEL1*, and *TINF2*.

Following the X-linked DC due to *DKC1* pathogenic variants, *TERT* and *TERC* variants were found in autosomal dominant DC, pulmonary fibrosis, and inherited aplastic anemia. The heterozygous variants lead to haploinsufficiency and impaired telomerase activity [9]. Biallelic variants are rarely found and are responsible for a more severe phenotype, such as the Hoyeraal-Hreidarsson syndrome or, in the classical DC, in an autosomal recessive pattern [33]. In addition, the inheritance pattern may be variable for the same gene among telomeropathies: *RTEL1* and *NHP2* are autosomal recessive in DC but autosomal dominant in familial pulmonary fibrosis [7, 55, 56]. In this scenario, it is reasonable to speculate that the heterozygosity of specific genes may be linked to a disease phenotype that only manifests later in life.

A group of premature aging diseases overlap symptoms with telomeropathies and are caused by pathogenic variants in DNA repair proteins that contribute to telomere preservation. These syndromes, such as RECQ helicase disorders and progeria, are called "secondary telomeropathies" [45]. On the other hand, patients with DC and other telomeropathies may not be classified as exhibiting an aging syndrome. Apart from premature hair greying, an important feature in many telomeropathy patients, they do not appear older than their age, and there is no increased incidence of age-related diseases such as atherosclerosis or Alzheimer's within this group [9]. Thus, it is a mistake to categorically attribute telomere shortening as the sole cause of aging, at most, it is one of the contributing factors.

Excessive telomere shortening in telomeropathy patients does increase their risk of developing cancer. DC patients have a four-fold higher incidence of cancer when compared with the general population; the majority of them are hematological malignancies. For example, for acute myeloid leukemia, there is a 24-fold greater incidence than in healthy individuals [1].

Not just short but excessively long telomeres increase the risk of cancer through distinct yet partially overlapping mechanisms. Genome-wide association studies (GWAS) found a correlation of variants that lengthen telomeres with cancer risk, such as melanoma and leukemia [36]. Cells with longer telomeres can undergo more divisions; although stochastic, the probability of acquiring a driver mutation in an oncogene exponentially increases at each round of cell division [28].

On the other hand, the "telomere paradox" links short-for-age telomeres to an increased risk of developing cancer [41], as observed in telomeropathies. Telomere erosion may be a tumor-suppressive mechanism, signaling DNA damage response and, ultimately, cell death. However, when these pathways fail as a consequence of driver or additional mutations (e.g., in *TP53*), the genetic instability that follows may elicit ALT mechanisms or telomerase reactivation. Interestingly, these two mechanisms of telomere maintenance are mutually exclusive in cancer [8].

The mechanisms underlying telomerase reactivation in most human cancers rely on *TERT* upregulation through point mutations in its coding or promoter regions [14]. While *TERC* is ubiquitously expressed in all normal human cells, TERT is not detectable in most of these cells but instead expressed in 85–90% of tumor cells [26]. Thus, the telomerase catalytic subunit is an attractive target for inhibition in most cancers. However, attempts in this way were unsuccessful for two reasons. First, some normal cells transiently express telomerase for tissue renewal, as the hematopoietic stem and progenitor cells, and such a therapy could give rise to a dangerous adverse effect. Second, there is a period between telomerase inhibition and telomere crisis to trigger apoptosis, which may be too long in cancer therapy [50].

A Short Talk on Telomere Measurement

Evaluating telomere length serves a purpose beyond merely satisfying one's curiosity about its "cellular age": it holds significant value in various diagnostic scenarios. When diagnosing human patients suspected of telomeropathy, measuring telomere length often represents one of the initial steps. Telomere length is typically assessed in peripheral blood leukocytes and can be expressed as an absolute measurement in kilobases (kb) or as a relative value. Given that telomeres naturally erode with age in the healthy population, it is essential to adjust telomere measurements for age when classifying a patient's telomere as normal or short. This classification is based on telomere length curves (percentiles) derived from healthy individuals: short telomeres are below the tenth percentile, while very short telomeres are those below the first percentile (Fig. 3.1, lower graph).

A human diploid cell has 46 chromosomes, thus 92 telomeres before entering the cell cycle, with a degree of variability in these telomere lengths. Ideally, methods that estimate individual telomere lengths, rather than just the average, are desirable, given that the shortest telomeres trigger senescent cell growth arrest [24, 28]. However, most currently validated methods give the telomere length as an average from cells or DNA samples. Four main methods are applied to assess telomere length, each with strengths and limitations.

The Southern blot analysis of terminal restriction fragment length is the gold standard for quantitation of mean telomere length, gives values in kb, and is reproducible [21]; however, it is a low-throughput and time-consuming assay and has limitations when telomeres are very short. The Single Telomere Length Analysis (STELA) is a PCR-based approach followed by Southern blot detection that assesses the full spectrum of telomere lengths from individual chromosomes and is accurate for detection of very short telomeres [4]. However, the method is not validated for clinical use.

A cheaper and faster method is based on quantitative PCR. The relative telomere length is given as a ratio of the signal produced by primers that anneal to the telomeres (T) and to a housekeeping gene known to have only one copy in the genome (S), producing the T/S ratio [11]. The intra- and inter-center variability in the results is high, limiting the method to research purposes.

Finally, the combination of flow cytometry with fluorescence in situ hybridization (flow-FISH) to detect telomeres is a widely used method due to its robustness, reproducibility, and validation for diagnosis [3]. The method has a good agreement and correlation compared to the standard technique Southern blot [20], but is relatively expensive and requires living cells.

The Termini Part of This Chapter

"Cap" this in mind: telomere homeostasis is crucial for overall genome stability as telomeres protect the linear chromosome ends from degradation and rearrangements. In most human somatic cells, telomere erosion is a natural hallmark of aging, thus limiting cellular lifespan. However, excessively short telomeres not just boost the aging process but also impair continued tissue regeneration and may be translated into severe diseases. On the other hand, excessive telomere elongation is not a synonym for longevity; instead, it is a dangerous situation also associated with cancer development.

The introduction of the telomerase catalytic component into telomerase-silent human cells is sufficient to bypass replicative senescence in the majority, but not all cases [6], which highlights that cell immortalization is not so obvious for manipulation. The mythical human idea of pursuing the Fountain of Youth remains unsolved, as telomere manipulation (artificial elongation) is not safe. A myriad of cellular and molecular pathways not completely elucidated are tightly regulated to sustain the gears of this biological clock.

References

1. Alter B. P., Giri N., Savage S. A. et al. (2018). Cancer in the National Cancer Institute inherited bone marrow failure syndrome cohort after fifteen years of follow-up. Haematologica 103(1):30–39.
2. Azzalin, C. M. et al. (2007). Telomeric repeat-containing RNA and RNA surveillance factors at mammalian chromosome ends. Science 318(5851), 798–801.
3. Baerlocher G. M., Vulto I., de Jong G., Lansdorp P. (2006). Flow cytometry and FISH to measure the average length of telomeres (flow FISH). Nat Protoc 1(5): 2365–2376.
4. Baird D. M., Rowson J., Wynford-Thomas D., Kipling D. (2003). Extensive allelic variation and ultrashort telomeres in senescent human cells. Nat Genet 33(2):203–7.
5. Blackburn, E. H., C. W. Greider, and J. W. Szostak (2006). Telomeres and telomerase: the path from maize, Tetrahymena and yeast to human cancer and aging: Nat Med 12, 1133–1138.
6. Bodnar A. G., Ouellette M., Frolkis M., et al. (1998). Extension of lifespan by introduction of telomerase into normal human cells. Science; 279(5349):349–352.
7. Borie R., Kannengiesser C., Antoniou K., Bonella F., Crestani B., et al. (2023). European Respiratory Society statement on familial pulmonary fibrosis. Eur Respir J; 61(3):2201383.
8. Brennan, C. W., R. G. Verhaak, A. McKenna, B. Campos, et al. (2013) The somatic genomic landscape of glioblastoma. Cell; 155:462–477.
9. Calado, R. T., and Young N. S. (2009). Telomere diseases: N Engl J Med, v. 361, p. 2353–65.

10. Calado, R. T., Yewdell W., Wilkerson K., Regal J., Kajigaya S., Stratakis C, Young N. S. (2009). Sex hormones, acting on the *TERT* gene, increase telomerase activity in human primary hematopoietic cells. *Blood*; 114(11): 2236–2243.
11. Cawthon R. M. (2002). Telomere measurement by quantitative PCR. Nucleic Acids Research. 30 (10): e47.
12. Cawthon R. M., Smith K. R., O'Brien E., Sivatchenko A., Kerber R. A. (2003). Association between telomere length in blood and mortality in people aged 60 years or older. Lancet. 361(9355): 393–395.
13. Cesare, A. J., and R. R. Reddel (2010). Alternative lengthening of telomeres: models, mechanisms and implications: Nat Rev Genet, v. 11, p. 319–30.
14. Chiba K., Lorbeer F. K., Shain A. H., et al. (2017). Mutations in the promoter of the telomerase gene TERT contribute to tumorigenesis by a two-step mechanism. Science; 357(6358):1416–1420.
15. Cong, Y. S., Wright, W. E., and Shay, J. W. (2002). Human telomerase and its regulation. Microbiol Mol Biol Rev 66, 407–25.
16. Cusanelli E., Romero C. P., Chartrand P. (2013). Telomeric non-coding RNA TERRA is induced by telomere shortening to nucleate telomerase molecules at short telomeres. Molecular Cell; 51(6):780–791.
17. de Lange, T. (2018). Shelterin-Mediated Telomere Protection. Annual Review of Genetics 52:1, 223–247.
18. Egan, E. D., and K. Collins (2010). Specificity and stoichiometry of subunit interactions in the human telomerase holoenzyme assembled in vivo: Mol Cell Biol, v. 30, p. 2775–86.
19. Griffith JD, Comeau L, Rosenfield S, et al. (1999). Mammalian telomeres end in a large duplex loop. Cell; 97(4):503–514.
20. Gutierrez-Rodrigues F, Santana-Lemos BA, Scheucher PS, et al. (2014). Direct comparison of flow- FISH and qPCR as diagnostic tests for telomere length measurement in humans. PloS One; 9(11):e113747.
21. Harley, C. B., Futcher, A. B., and Greider, C. W. (1990). Telomeres shorten during ageing of human fibroblasts. Nature 345, 458–60.
22. Hayflick, L. (1965). The limited in vitro lifetime of human diploid cell strains. Exp Cell Res 37, 614–36.
23. Heiss, N. S., Knight, S. W., Vulliamy, T. J., Klauck, S. M., Wiemann, S., Mason, P. J., Poustka, A., and Dokal, I. (1998). X-linked dyskeratosis congenita is caused by mutations in a highly conserved gene with putative nucleolar functions. Nat Genet 19, 32–8.
24. Hemann M. T., Strong M. A., Hao L. Y., Greider C. W. (2001). The shortest telomere, not average telomere length, is critical for cell viability and chromosome stability. Cell; 107(1): 67–77.
25. Herbig, U., Jobling, W. A., Chen, B. P., Chen, D. J., and Sedivy, J. M. (2004). Telomere shortening triggers senescence of human cells through a pathway involving ATM, p53, and p21(CIP1), but not p16(INK4a). Mol Cell 14, 501–13.
26. Kim N. W., Piatyszek M. A., Prowse K. R., et al. (1994). Specific association of human telomerase activity with immortal cells and cancer. Science; 266(5193):2011-2015.
27. Kirwan, M., and Dokal, I. (2009). Dyskeratosis congenita, stem cells and telomeres. Biochim Biophys Acta 1792, 371–9.
28. Lansdorp P. M. (2022). Telomeres, aging, and cancer: the big picture. Blood; 139(6):813–821.
29. Le Guen T, Jullien L, Touzot F, et al. (2013). Human RTEL1 deficiency causes Hoyeraal-Hreidarsson syndrome with short telomeres and genome instability. Hum Mol Genet, 22(16):3239–49.
30. Li JS, Miralles Fuste J, Simavorian T, Bartocci C, Tsai J, Karlseder J, Lazzerini Denchi E. (2017). TZAP: A telomere-associated protein involved in telomere length control. Science; 355: 638–641.
31. Makarov, V. L., Hirose, Y., and Langmore, J. P. (1997). Long G tails at both ends of human chromosomes suggest a C strand degradation mechanism for telomere shortening. Cell 88, 657–66.

32. Markiewicz-Potoczny M, Lobanova A, Loeb AM, et al. (2021). TRF2-mediated telomere protection is dispensable in pluripotent stem cells. Nature; 589(7840):110–115.
33. Marrone A., Walne A., Tamary H., Masunari Y., Kirwan M., Beswick R., Vulliamy T., Dokal I. (2007). Telomerase reverse-transcriptase homozygous mutations in autosomal recessive dyskeratosis congenita and Hoyeraal-Hreidarsson syndrome. Blood; 110(13):4198–205.
34. Martens U. M., Chavez E. A., Poon S. S., Schmoor C., Lansdorp P. (2000). Accumulation of short telomeres in human fibroblasts prior to replicative senescence. Exp Cell Res; 256(1):291–299.
35. McClintock, B. (1941). The stability of broken ends of chromosomes in *Zea mays*. Genetics 26, 234–82.
36. McNally E., Luncsford J., Armanios M. (2019). Long telomeres and cancer risk: the price of cellular immortality. J Clin Invest; 129(9):3474–3481.
37. Mitchell, J. R., Wood, E., and Collins, K. (1999). A telomerase component is defective in the human disease dyskeratosis congenita. Nature 402, 551–5.
38. Moon DH, Segal M, Boyraz B, et al. (2015). Poly(A)-specific ribonuclease (PARN) mediates 3′-end maturation of the telomerase RNA component. Nat Genet, 47(12):1482–8.
39. Moyzis, R. K., Buckingham, J. M., Cram, L. S., Dani, M., Deaven, L. L., Jones, M. D., Meyne, J., Ratliff, R. L., and Wu, J. R. (1988). A highly conserved repetitive DNA sequence, (TTAGGG)n, present at the telomeres of human chromosomes. Proc Natl Acad Sci U S A 85, 6622–6.
40. Muller, H. (1938). The remaking of chromosomes. Collect. Net, 13, pp. 181–198.
41. Nassour J, Schmidt TT, Karlseder J. (2021). Telomeres and cancer: resolving the paradox. Annu Rev Cancer Biol; 5(1): 59–77.
42. Nergadze S. G., Farnung B. O., Wischnewski H., et al. (2009). CpG-island promoters drive transcription of human telomeres. RNA; 15(12): 2186–2194.
43. Niewisch M. R., Savage S. A. (2019). An update on the biology and management of dyskeratosis congenita and related telomere biology disorders. Expert Rev Hematol;12(12): 1037–1052.
44. Olovnikov, A. M. (1971). Principle of marginotomy in template synthesis of polynucleotides. Dokl Akad Nauk SSSR 201, 1496–9.
45. Opresko P., Shay J. W. (2017). Telomere-associated aging disorders. Ageing Res Rev; 33:52–66.
46. Penev A, Bazley A, Shen M, Boeke JD, Savage SA, Sfeir A. (2021). Alternative splicing is a developmental switch for hTERT expression. Mol Cell; 81(11):2349–2360.e6.
47. Redon S., Reichenbach P., Lingner J. (2010). The non-coding RNA TERRA is a natural ligand and direct inhibitor of human telomerase. Nucleic Acids Res; 38(17):5797–806.
48. Schumacher B., Pothof J., Vijg J, Hoeijmakers J. (2021). The central role of DNA damage in the ageing process. Nature;592(7856): 695–703.
49. Shay, J. W., and Wright, W. E. (2000). Hayflick, his limit, and cellular ageing. Nat Rev Mol Cell Biol 1, 72–6.
50. Shay, J. W., and Wright, W. E. (2006). Telomerase therapeutics for cancer: challenges and new directions. Nat Rev Drug Discov; 5(7):577–84.
51. Stansel R. M., de Lange T., Griffith J. D. (2001). T-loop assembly in vitro involves binding of TRF2 near the 3′ telomeric overhang. EMBO J. 20:5532-40.
52. Uringa, E. J., Youds, J. L., Lisaingo, K., Lansdorp, P. M., and Boulton, S. J. (2011). RTEL1: an essential helicase for telomere maintenance and the regulation of homologous recombination. Nucleic Acids Res 39, 1647–55.
53. Vannier JB, Sarek G, Boulton SJ. (2014). RTEL1: functions of a disease-associated helicase. Trends Cell Biol; 7, 416–25.
54. Venteicher, A. S., and S. E. Artandi (2009). TCAB1: driving telomerase to Cajal bodies: Cell Cycle, v. 8, p. 1329–31.
55. Vulliamy T., Beswick R., Kirwan M., Marrone A., Digweed M., Walne A., Dokal I. (2008). Mutations in the telomerase component NHP2 cause the premature ageing syndrome dyskeratosis congenita. Proc Natl Acad Sci U S A;105(23):8073–8.

56. Walne, A. J., Vulliamy, T., Kirwan, M., Plagnol, V., and Dokal, I. (2013). Constitutional mutations in RTEL1 cause severe dyskeratosis congenita. Am J Hum Genet 92, 448–53.
57. Zinsser F. (1906) Atrophia cutis reticularis cum pigmentatione dystrophia unguium et leukoplakia oris; poikiloderma atrophicans vascularis Jacobi. Inkonographia Dermatologia; 5:219–223.

Chapter 4
Epigenetic Changes in Aging: The Contribution of SIRT1 to Longevity

Débora Kristina Alves-Fernandes and Miriam Galvonas Jasiulionis

Introduction

Aging represents the gradual and innate progression of physical, psychological, and functional alterations observed in living organisms over time. This intricate biological phenomenon is influenced by a convergence of genetic, environmental, and lifestyle factors. As a rule, aging is associated with the deterioration of multiple bodily functions and an elevated susceptibility to age-related diseases. Numerous in vitro and in vivo studies have demonstrated the crucial role of cellular senescence in aging process [1–4]. However, there are also contrasting studies suggesting the cellular senescence may be independent of aging [5, 6], which has led to ongoing debate and controversy in the field [7]. It is important to note that role of senescence in aging appears to be contingent upon the specific cell types and tissues involved [7].

It becomes evident that as aged organisms, there is a general reduction in histones, accompanied by local and global chromatin remodeling, which also results in an imbalance between activating and repressive histone modifications, leading to global transcriptional changes observed across various aging models [8]. Moreover, especially in mammalian systems, it has been observed global and local changes in DNA methylation, site-specific loss and acquisition of heterochromatin, as well as substantial nuclear reorganization [8].

D. K. Alves-Fernandes · M. G. Jasiulionis (✉)
Department of Pharmacology, Escola Paulista de Medicina, Universidade Federal de São Paulo, São Paulo, SP 04023-900, Brazil
e-mail: mgjasiulionis@unifesp.br

V. Bueno (ed.), *Cellular and Molecular Aspects of Ageing*,
https://doi.org/10.1007/978-3-031-55022-5_4

Epigenetic Changes in Aging

In recent years, the observation of age-dependent changes in epigenetic regulation have suggested that epigenetic mechanisms have a crucial role in the aging process [7].

Epigenetic events are reversible and heritable mechanisms that involve chromatin modifications, leading to altered gene expression without any fundamental change in DNA sequences. Eukaryotic chromatin is a densely compacted structure composed of repeating structural units known as nucleosomes. Each nucleosome comprises a histone octamer, consisting of two copies of each core histone (H2A, H2B, H3 and H4), along with histone variants like H3.3, macroH2A, and H2A.Z. This histone octamer is wrapped around 147 base pairs of DNA [9–11]. H1 histones bind both the nucleosome and the linker DNA, which spans across adjacent nucleosome [12]. Epigenetic factors and chromatin conformation have influence in gene expression over extended time scales, persisting through cell divisions, by modulating the accessibility of transcriptional machinery to DNA [8].

While SIRT1 is primarily recognized as a histone deacetylase, it serves as a pivotal epigenetic regulator involved not only in histone deacetylase but also in the recruitment and activation of DNA and histone methyltransferases [13]. However, in this first section we will narrow our focus to epigenetic marks associated with aging that are not directly linked to SIRT1.

DNA Methylation

In 1967, the DNA methylation was firstly related to the age in spawning salmon [14]. Subsequent studies have showed that DNA methylation experiences a downregulation in numerous tissues, both in mice and humans, as they undergo the aging process [15–18]. DNA methylation involves the addition of methyl groups (CH3) to DNA molecule by DNA methyltransferases (DNMTs), namely DNMT1, DNMT3A and DNMT3B [19–21]. This modification primarily occurs at the fifth position of the cytosines ring within phospho-guanine (CpG) dinucleotides, resulting in the formation of 5-methylcytosine (5-mC). Remarkably, in the mammalian genome, approximately 60–90% of CpG sites undergo this methylation process [22]. While it is widely acknowledged that DNA methylation at promoter regions is associated with gene silencing, further research is needed to fully explore the role of gene body methylation [11, 23]. DNMT1 expression decreases with age, leading to a decline in DNA methylation levels. Interestingly, DNMT1 mutants, which are responsible for the degeneration of specific central and peripheral neurons, have been observed to relocate to the cytoplasm and aggregate into aggresomes, failing to bind to heterochromatin [24]. Conversely, the expression of DNMT3A and DNMT3B increase according to age, and these DNA methylases play a pivotal role in the de novo methylation of CpG islands in mammalian cells [25, 26]. DNMT3A and DNMT3B have function alone or in complex with DNMT3L [19]. The family of Ten-eleven translocation (TET) proteins, TET1, TET2 and TET3, plays a central role in the removal of DNA methylation [27–29]. Mutations in TET2 and DNMT3A genes among elderly individuals have been linked to heightened expression of pro-inflammatory cytokines and the development of chronic inflammation, which, in turn, is closely associated with conventional cardiovascular disease (CVD) [22, 30].

Significantly, there is a heightened variability in DNA methylation patterns as individuals age, referred to as age-associated variably methylated positions (aVMPs) [31, 32]. This phenomenon was initially discovered through twin studies, where older monozygotic twins exhibited a greater degree of methylation variability in their overall 5-mC content compared to their younger counterparts [32]. Alongside aVMPs, there are also distinct CpG sites referred to as age-associated differentially methylated positions (aDMPs) [33], which exhibit a decrease with age across six mammalian species, like as human, mice, dogs, naked mole rats, rhesus macaques, and humpback whales [31]. As such, the shift in DNA methylation is linked to various CpG sites, encompassing aVMPs and aDMPs, which can be quantified to evaluate epigenetic age [22]. Notably, the understanding of the significance of DNA methylation patterns during aging has been significantly advanced by two large-scale studies, that exhibited differential methylation with age. In these studies, the authors characterized 353 and 71 CpG sites, respectively, which can serve as aging predictors in human tissue [34–36].

Analysis of DNA methylomes between newborns and centenarians people revealed that DNA hypomethylation predominantly occurs in CpG-poor promoters or tissue-specific promoters [17]. Furthermore, hypomethylated differentially methylated regions (DMRs) is more commonly observed in genomics regions enriched with Polycomb proteins/regulatory proteins (e.g., EZH2) or permissive histone modifications (methylation and acetylation) [37]. Conversely, DNA hypermethylation is identified primarily in CpG-rich sequences [38]. The transition toward hypomethylation of constitutive heterochromatin and hypermethylation of promoters in cell-cycle-promoting genes, age-dependent alterations in DNA methylation may instigate heightened genomic instability and culminate in a persistent cell-cycle arrest. Moreover, the varying methylation patterns of promoter CpGs may also play a role in the observed alterations in gene expression profiles associated with senescence and aging [7].

Histone Modifications

Numerous post-translational histone modifications play a pivotal role in either activating or silencing gene expression, thereby intricately regulating the aging process. These modifications encompass methylation, acetylation, phosphorylation, ubiquitination, among others [39].

In general, acetylation and phosphorylation of histone tails are commonly associated with euchromatin, while histone methylation is predominantly enriched in heterochromatin regions [40]. Here our focus will be on histone methylation and acetylation at lysine residues, which are extensively studied and are known to be closely associated with aging process.

Histone Methylation

Histone methylation can take place on all basic amino acid residues: arginines [41], histidines [42] and lysines [43]. Arginines can be mono-methylated (me1) [41], symmetrically (me2s), or asymmetrically (me2a) dimethylated on their guanidinyl group [44]. Although monomethylation of histidines [44, 45] has been reported, it appears to be rare and lacks further characterization [46]. Lysines, the most studied amino acid residue, can undergo mono-methylation (me1) [43], di-methylation (me2) [47], or tri-methylation (me3) on their ε-amine group [48]. In general, previous research suggests that the methylation of H3K4, H3K36, and H3K79 enhances transcription, while methylation of H3K9, H3K27, and H4K20 has a suppressive effect on transcription levels [46].

The deletion of RPH1, responsible to demethylate H3K36 me2/3, resulted in increased H3K36me3 levels in *S. cerevisiae*, consequently extending the lifespan of the yeast [49]. By studying the genome-wide pattern of H3K36me3 in somatic cells of young and old *C. elegans*, it was observed that genes exhibiting significant expression changes during aging are characterized by low or even undetectable levels of H3K36me3 within their gene bodies regardless of their corresponding mRNA abundance [50]. Similar results were observed in *Drosophila melanogaster*, where the inactivation of methyltransferase *met-1* led to a reduction in global H3K36me3 marks. This interference in mRNA expression changes associated with aging resulted in a shorter lifespan [50]. H3K36me2 was also shown to be relevant in the aging process. When the H3K36 dimethyltransferase SET18 was deleted in *C. elegans*, the lifespan was extended in a pattern dependent of DAF16 [51]. DAF-16, the sole ortholog of the forkhead transcription factor (FOXO) in *C. elegans*, undergoes phosphorylation by PI3K-AKT1/2, SGK-1 and PDK-1 signaling pathways [52–54]. Phosphorylated DAF-16 is inhibited from entering the nuclei and activating its target genes, which include genes related to antiaging and stress-resistance [55–58]. While they serve different functions, these discoveries collectively confirm the significance of di- and trimethylation of H3K36 in promoting longevity [11].

Trimethylation of lysine 27 on histone H3 (H3K27me3) is an epigenetic mark associated with repression and has been observed to decrease during aging in *C. elegans*. The regulation of H3K27me3 involves the Polycomb Repressive Complex 2 (PRC2) and its reversal the Ubiquitously transcribed TPR on X1 (UTRX1). Notably, the lifespan was extended through the depletion of the demethylase UTRX1, leading to an increase of H3K27me3 levels [23]. Interestingly, other studies have reported an unexpected increase in H3K27me3 abundance during aging in different species. For example, depleting homologs of PRC2 members in *Drosophila* led to decreased levels of H3K27me3 and an increase in longevity [59].

The H3K9me3 and H4K20me3 histone marks, which are also considered repressive marks involved with aging process, have an indirect association with SIRT1 and will be addressed in a later discussion.

Histone Acetylation

The chromatin access is highly dependent of enzymes that modify the histone core, as histone acetylases (HAT) and deacetylases (HDACs). Alteration in the gene expression profile related with aging by altering histone acetylation may be crucial for longevity manipulation [60]. Histone acetylation has long been implicated in aging, both in Yeast and Fly. The overall model suggests that histone hypoacetylation prolongs lifespan by promoting autophagy while suppressing oxidative stress and necrosis [61]. Although this effect is achieved through various mechanisms, such as the inhibition of acetyl-CoA producing enzymes, spermidine supplementation, or the inactivation of HATs, it is not yet fully understood [62]. In IMR90 cells, p300 HAT, a factor that promotes the formation of active enhancer elements in the non-coding genome, significantly elevates the levels of H3K122ac and H3K27ac at the proximal promoters of senescence-specific genes. P300 has been confirmed as a primary driver of senescent phenotypes, and depleting P300 alone is sufficient to downregulate senescence genes and delay replicative senescence [62]. Furthermore, in *C. elegans*, CBP-1, the ortholog of the mammalian CBP/P300 HAT, is an essential regulator of unfolded protein response and mediates H3K27ac and H3K18ac in response to mitochondrial stress [63]. In addition, the CBP-1 knockdown is associated to decreased lifespan in a manner dependent of heat shock transcription factor (HSF-1) [64].

In contrast to HATs that is linked to aging, the HDACs have been associated to longevity. KAT7 inactivation leads to a decrease in H3K14ac level, which in turn mitigates senescence in human mesenchymal precursor cell (hMPC) [65]. Similarly, H3K18ac and H3K56ac serve as negative senescence markers in flies and yeast, respectively [66, 67]. In aged yeast, the level of H3K56ac decreases, while H4K16ac level increases, resulting in the silencing of telomeric repeats, but this mark will be discussed latter [68]. Although by different mechanism of action, the knockdown of H3K56ac and genes encoding histone deacetylases Hst3 and Hst4 during yeast aging led to shorten lifespan [69]. These studies suggest that preserving the fundamental structure is crucial for slowing down the aging process, and reveal that histone deacetylases play an essential role in extending lifespan.

Involvement of SIRT1 with Aging and Longevity

Sir2 protein was the first sirtuin discovered in yeast, but the SIR2 family is conserved from bacteria to mammals [70, 71]. Evidences showed that SIR2 was related with silencing, recombination suppression and lifespan extension in vivo [72].

Mammalian sirtuins (SIRT1 to SIRT7), members of the NAD+-dependent class III HDACs, exhibit distinct target profiles, cellular localization, and functional activities. SIRT1, SIRT6, and SIRT7 localize predominantly in the nucleus, actively participating in the epigenetic regulation by primarily targeting histones and transcription factors. In contrast, SIRT3, SIRT4, and SIRT5 are found within the mitochondria, with SIRT3, in particular, recognized for safeguarding mitochondrial DNA and inhibiting apoptosis through epigenetic mechanisms [73–76]. SIRT1, SIRT2 and SIRT3 may be shuttle between nucleus and cytoplasm depending of cellular stimulus [73, 74, 77–79]. This suggests that their functional activities are closely tied to their cellular localizations.

SIRT1, an ortholog of Sir2 and the most extensively studied sirtuin, is recognized as a master epigenetic regulator intricately involved in numerous crucial biological processes, encompassing DNA response, DNA repair, genomic stability maintenance, cellular reprograming, metabolism, apoptosis, aging, and tumorigenesis [13, 80–82]. SIRT1 exhibits a wide range of targets, including both histone and non-histone proteins, and through deacetylation, it can either activate or inhibit these proteins. In this chapter, we will delve into the primary epigenetic alterations associated with aging, with a specific focus on the pivotal role of SIRT1. We aim to demonstrate how the absence of this master regulator significantly contributes to the aging process and to pursuit the longevity. The primary epigenetic alterations associated with aging, potentially involving SIRT1, are depicted in Fig. 4.1.

SIRT1 and DNA Methylation

Numerous studies showed that SIRT1 is downregulated in aged organisms in human and mouse tissues, including liver, heart, kidney, brain and lung [83–85]. Evidences has shown that in aged renal samples, SIRT1 promotor region may be methylated by the renal progenitor transcription factor SIX2, resulting in transcriptional downregulation of SIRT1 and its downstream targets AKT and GSK3β during aging [86].

It was demonstrated that SIRT1 has effects on DNA methylation across the genome and notably influences the expression of genes responsive to dietary restriction [87]. Dietary restriction has been shown to increase lifespan in evolutionarily distinct species, ranging from yeast to rodents [88].

Besides that, SIRT1 has pleiotropic effects on DNA methylation at Polycomb group proteins target genes (PCGTs), which in turn contribute to healthspan. In stem cells, PCGTs are repressed by epigenetic mechanisms, partly through the action of

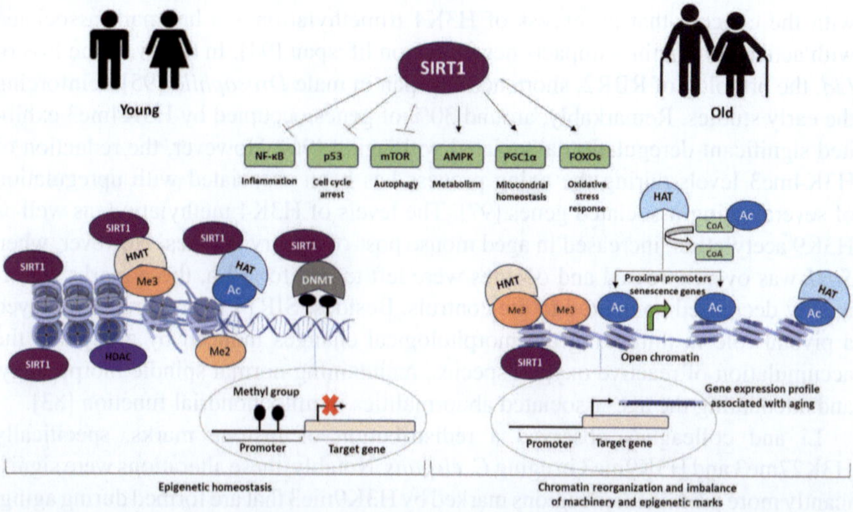

Fig. 4.1 Overview of epigenetic changes in aging and the role of SIRT1 in longevity. In this figure, we illustrate the primary epigenetic alterations that occur during the aging process and how SIRT1 plays a crucial role in promoting longevity. In young individuals, there is a state of epigenetic homeostasis. The equilibrium is maintained by precise control over the epigenetic machinery, resulting in a harmonious balance among histone modifications, DNA methylation and chromatin formation. These factors collectively contribute to a uniform gene expression profile across different cell types. SIRT1 expression is notably elevated in young individuals, granting it a multifaceted role beyond its deacetylase function. SIRT1 actively participates in heterochromatin formation and activation of key enzymes in the epigenetic machinery. Additionally, SIRT1 influences the inhibition or activation of essential proteins associated with the aging process, highlighting its significance for promoting longevity. As individuals age, a variety of exogenous and endogenous factors come into play, leading to significant alterations in epigenetic information. These changes create an abnormal chromatin state characterized by the loss of histones, decreased DNA methylation, and modifications to critical epigenetic marks. These alterations collectively contribute to changes in chromatin structure, resulting in an altered gene expression profile that is associated with aging phenotype

Polycomb Repressive Complexes (PRCs). These genes have previously been shown to correspond to loci that are particularly susceptible to age-related changes in DNA methylation [89].

SIRT1 and Histone Methylation

While regulators of histone methylation have been linked to aging in *C. elegans* [90–93] and *Drosophila* [59], the characterization of their role and mechanisms of action remains limited [94]. Deficiencies in members of the ASH-2 complex (ASH-2 itself, WDR-5 and H3K4 methyltransferase SET2) responsible for trimethylation of lysine 4 of H3 (H3K4me3) are related to extend lifespan in *C. elegans.* In contrast, the H3K4 demethylase RBR-2 is essential for maintaining a normal lifespan, aligning with the concept that an excess of H3K4 trimethylation – a hallmark associated with active chromatin—impacts negatively on lifespan [94]. In this way, the loss of *Lid*, the ortholog of RBR2, shortened lifespan in male *Drosophila* [95], reinforcing the early studies. Remarkably, around 30% of genes occupied by H3K4me3 exhibited significant deregulation associated with aging [96]. However, the reduction of H3K4me3 levels during the aging process has been associated with upregulation of several aging-associated genes [97]. The levels of H3K4 methylation, as well as H3K9 acetylation, increased in aged mouse post-ovulatory oocytes. However, when *Sirt1* was overexpressed and oocytes were left to age for 12 h, these marks significantly decreased compared to the controls. Besides, SIRT1 overexpression played a pivotal role in mitigating the morphological changes induced by aging and the accumulation of reactive oxygen species, maintaining normal spindle morphology, and attenuating the age-associated abnormalities in mitochondrial function [83].

Li and colleagues observed a redistribution of histone marks, specifically H3K27me3 and H3K9me3 in aging *C. elegans*. Notably, these alterations were significantly more prominent in regions marked by H3K9me3 that are formed during aging, and these changes were not evident in developmental worms. These regions, known as aging-specific repressive regions (ASRRs), are found in genic regions that exhibit high levels of H3K9me2 and H3K36me2 during larval stages. Corroborating these results, the maintenance of elevated H3K9me2 levels in these regions has been shown to correlate with a longer lifespan [98]. In *Drosophila*, the depletion of KDM4A, a H3K9me3 demethylase, resulted in the downregulation of factors associated with the male sex-determination gene, as well as several genes linked to longevity [99]. Consequently, the expression of the methyltransferase SUV39H1, responsible for H3K9me3, was observed to decrease during aging of both human and murine cells [100]. Notably, SIRT1 has the remarkable capability to recruit and directly activate the methyltransferase SUV39H1 through its deacetylase activity, as demonstrated by VAQUERO and colleagues (2007). Consequently, SIRT1 triggers the generation of H3K9me3, an epigenetic mark intricately linked to the formation of heterochromatin [101].

An enrichment of H4K20me3 was observed in animals older than 30 days when comparing kidney and liver samples from rats with 10, 30, 300 and 450 days of age. On the other hand, mono and dimethylation of H3K20 did not exhibit changes associated with aging process [102]. On the contrary, a decrease in H4K20me3 levels was observed in mouse embryogenic fibroblasts (MEFs), which contributed to cellular senescence and cardiac aging. This reduction served as a responsive epigenetic target of canonical TGF-β signaling, being negatively regulated by miR-29. Therefore, it appears that the alteration in H3K20me3 level is cell type-specific and dependent on TGF-β signaling [103]. It was demonstrated that SUV39H1 may allosterically methylate SET8, which stimulates the increase in H4K20me3 level [104]. These findings highlight their potential role in promoting longevity.

SIRT1 and Histone Acetylation

As early mentioned, SIRT1 is a histone deacetylase that plays a role in chromatin formation. It is responsible for deacetylating specific marks, including H3K9, H3K79, and preferentially H4K16, as demonstrated by VAQUERO and coworkers. Additionally, SIRT1 has been shown to interact with and deacetylate H1K26 in in vitro studies, contributing to the heterochromatin formation [105].

In yeast, Sir2, an ortholog of SIRT1 in mammals, has been found to decrease in replicatively old yeast cells. This coincides with an increase in H4K16ac level and loss of histones at specific subtelomeric regions, leading to compromised transcriptional silencing at these loci [68]. The acetylation of H4K16 disrupts nucleosome interactions and triggers chromatin unfolding, which is an indicative of active transcription [106]. In opposition to, deacetylated H4K16 is usually associated with heterochromatin and transcriptional silencing [107]. SIRT1, SIRT2 and SIRT3 have similar specificity for H4K16ac and H3K9ac. These enzymes together with HDAC2 are known to have deacetylase activity in H4K16 in high eukaryotes [107, 108].

Investigating livers isolated from 6, 15 and 30 months-old rats, it was observed a decrease in H3K9ac levels with age [109]. In addition to SIRT1, other sirtuins such as SIRT6 also possess deacetylase activity on H3K9. Depletion of SIRT6 resulted in H3K9-related changes, leading to the formation of abnormal telomere structures [110].

SIRT1 Contributing to Non-epigenetic Mechanisms Essential for the Aging Process

SIRT1 has been extensively linked to the aging process, and its overexpression has been shown to promote lifespan extension in various organisms, including yeast, flies, worms and mammals [111–115].

Intermediate components in pleiotropic cellular pathways and various key transcription factors, which are likely to have an impact on healthspan, serve as substrate for deacetylation by SIRT1. Notable among these substrates is PGC1α, which governs mitochondrial biogenesis, as well as NF-κB, AMPK, mTOR, FOXOs, p53 and numerous others [116, 117].

NF-κB, a prominent transcription factor, becomes activated in numerous conditions linked to inflammation and aging. These include diseases like arthritis, atherosclerosis, diabetes, Alzheimer's and Parkinson's diseases [117]. SIRT1 interacts with and negatively regulates NF-κB signaling pathway through the deacetylation of lysine residue 310 on the RelA/p65 subunit of NF-κB. Acetylation of p65 enhances the transcriptional activity of the NF-κB complex [118], whereas deacetylation of Lys310 exposes the RelA/p65 protein to methylation at Lys314 and Lys315. This, in turn, enhances its ubiquitination and degradation [119]. In sequence, NF-κB system suppresses SIRT1-mediated functions by inhibiting the SIRT1 downstream targets [116]. NF-κB and SIRT1 engage in an antagonistic crosstalk, influencing the regulation of inflammation, metabolic disorders, and ultimately, the aging process. NF-κB drives a pro-inflammatory phenotype with glycolytic metabolism, while SIRT1 promotes oxidative respiration and anti-inflammatory responses, facilitating the resolution of inflammation [116].

Adenosine 5'monophosphate-activated protein kinase (AMPK) and SIRT1 have a close crosstalk in regulating energy metabolism, inflammation and aging [120, 121]. SIRT1 deacetylates LKB1 at K48, inducing cytoplasmic localization and activating AMPK [122]. In turn, AMPK promotes NAD+ synthesis, which is required to SIRT1 activity [123]. The activation of AMPK enhances SIRT1 activity by increasing cellular NAD+ levels, leading to the deacetylation of peroxisome proliferator-activated receptor coactivator 1α (PGC1α), another target of SIRT1 [124]. PGC1α serves as a primary transcriptional coactivator for maintaining mitochondrial homeostasis, and SIRT1 plays a crucial role in regulating metabolic homeostasis and mitochondrial functions. SIRT1 enhances mitochondrial biogenesis and oxygen consumption by deacetylating PGC1α [117, 125]. Mitochondria are essential for regulating cell death, influencing cellular energy production, and impacting on longevity [126].

SIRT1 and mTOR also participate in regulating the aging process by maintaining autophagy function [127]. The inhibition of mTOR activity is associated with extended lifespan in various organisms [117]. SIRT1 interacts with tuberous sclerosis protein 2 (TSC2), a component of the mTOR inhibitory-complex upstream of mTORC1 and negatively regulates mTOR signaling in a TSC2-dependent manner [128]. SIRT1 blocks mTOR signaling, restoring autophagy impairment induced by oxidative stress, and consequently improving the survival rate of embryonic stem cells [129].

FOXOs, Forkhead transcription factors, a family of proteins that act as sensors in the insulin signaling pathway, are involved in various physiological processes, responding to oxidative stress, facilitating DNA damage repair, regulating autophagy, inducing cell cycle arrest and regulating organism longevity [117]. SIRT1 may deacetylate FOXO3, promoting cell cycle arrest and enhancing resistance to oxidative stress. However, SIRT1 can also inhibit FOXO3's capacity of inducing cell death, thereby contributing to increased longevity [130].

Last but not least, the relationship between p53 and SIRT1 has been one of the best studied [82, 117]. The tumor suppressor P53 was one of the first non-histone substrates identified to be functionally associated with SIRT1's anti-senescence activity [131]. The acetylation of p53 promotes the activation of various genes involved with cell cycle arrest and apoptosis [117]. SIRT1 inhibits p53 by direct interaction via deacetylation at C-terminal lysine-382 residue [132]. The deacetylated p53 reduces its ability in regulating the transcription of p21, a cell cycle inhibitor. This reduction allows cells to re-enter the cell cycle after successful DNA repair [133].

Concluding Remarks

SIRT1 is a multifaceted deacetylase and, as discussed in this chapter, plays numerous roles that could potentially influence the aging process. Its impact on aging is not easily boiled down to a single direct effect due to its pleiotropic effects.

SIRT1 has gained attention for its role as a regulator of longevity and its therapeutic potential in preventing and treating age-associated diseases, including cardiovascular diseases, diabetes and neurodegenerative conditions [117]. It's worth noting that SIRT1 experiences reduced expression and activity in aging organisms.

Furthermore, SIRT1 interacts with and deacetylates numerous enzymes involved in the epigenetic machinery, thereby affecting essential epigenetic marks and transcriptional landscape. These alterations are especially prominent during aging process. The decreased expression and activity of SIRT1, combined with histone reduction, leads to chromatin reorganization, resulting in a shift in the transcriptional profile that favors aging and age-related diseases.

In addition to its epigenetic roles, SIRT1 also interacts with several key proteins, including NF-κB, AMPK, PGC1α, m-TOR, FOXO's and p53. These proteins are associated with various aspects of the aging process, such as inflammation, metabolism, oxidative stress, cell cycle arrest, apoptosis, and autophagy. These signaling pathways are closely linked to aging, further underscoring the potential importance of SIRT1 in promoting longevity.

References

1. Huang, S., Risques, R. A., Martin, G. M., et al. Accelerated telomere shortening and replicative senescence in human fibroblasts overexpressing mutant and wild-type lamin A. Exp. Cell Res. 2008, 314, 82–91.
2. Herbig, U., Ferreira, M., Condel, L., et al. Cellular senescence in aging primates. Science. 2006, 311:1257.
3. Farr, J. N., Fraser, D. G., Wang, H., et al. Identification of senescent cells in the bone microenvironment. J. Bone Miner. Res. 2016, 31, 1920–1929.
4. Biran, A., Zada, L., Abou Karam, P., et al. Quantitative identification of senescent cells in aging and disease. Aging Cell. 2017, 16, 661–671.
5. Ng M. H., Aminuddin B. S., Hamizah S., et al. Correlation of donor age and telomerase activity with *in vitro* cell growth and replicative potential for dermal fibroblasts and keratinocytes. 2009. *J. Tissue Viability* 18, 109–116.
6. Cristofalo, V. J., Allen, R. G., Pignolo, R. J., et al. Relationship between donor age and the replicative lifespan of human cells in culture: a reevaluation. Proc. Natl. Acad. Sci. U.S.A. 1998, 95, 10614–10619.
7. Sidler C, Kovalchuk O, Kovalchuk I. Epigenetic Regulation of Cellular Senescence and Aging. Front Genet. 2017, Sep 26;8:138.
8. Sen P, Shah PP, Nativio R, et al. Epigentic mechanisms regulating longevity and aging. Cell. 2016. 11; 166 (4): 822–839.
9. Luger, K.; Mäder, A.W.; Richmond, R.K.; et al. Crystal structure of the nucleosome core particle at 2.8 A resolution. Nature. 1997, 389, 251–260.
10. Richmond TJ, Davey CA. The structure of DNA in the nucleosome core. Nature. 2003 May 8;423(6936):145–50.

11. Saul D, Kosinsky RL. Epigenetics of Aging and Aging-Associated Diseases. Int J Mol Sci. 2021 Jan 2;22(1):401.
12. Zhou BR, Feng H, Kale S, et al. Distinct Structures and Dynamics of Chromatosomes with Different Human Linker Histone Isoforms. Mol Cell. 2021, Jan 7;81(1):166–182.e6. https://doi.org/10.1016/j.molcel.2020.10.038.
13. Alves-Fernandes DK, Jasiulionis MG. The Role of SIRT1 on DNA Damage Response and Epigenetic Alterations in Cancer. Int J Mol Sci. 2019, Jun 28;20(13):3153.
14. Berdyshev GD, Korotaev GK, Boiarskikh GV, et al. [Nucleotide composition of DNA and RNA from somatic tissues of humpback and its changes during spawning]. Biokhimiia. 1967, Sep–Oct;32(5):988–93. Russian.
15. Wilson, V. L. & Jones, P. A. DNA methylation decreases in aging but not in immortal cells. Science. 1983, 220, 1055–1057.
16. Wilson, V. L., Smith, R. A., Ma, S. et al. Genomic 5-methyldeoxycytidine decreases with age. J. Biol. Chem. 1987 262, 9948–9951.
17. Heyn, H. et al. Distinct DNA methylomes of newborns and centenarians. Proc. Natl Acad. Sci. USA. 2012, 109, 10522–10527.
18. Seale, K. et al. Making sense of the ageing methylome. Nat. Rev. Genet. 2022, 23, 585–605
19. Schübeler D. Function and information content of DNA methylation. Nature. 2015, Jan 15;517(7534):321–6.
20. Li, Y. et al. Stella safeguards the oocyte methylome by preventing de novo methylation mediated by DNMT1. 2018, Nature 564, 136–140.
21. Adhikari, S. & Curtis, P. D. DNA methyltransferases and epigenetic regulation in bacteria. FEMS Microbiol. Rev. 2016, 40, 575–591.
22. Wang, K, Liu H, Hu Q, et al. Epigenetic regulation of aging: implications for interventions of aging and diseases. Signal Transduction and Targeted Therapy. 2022, 7:374.
23. Jin B, Li Y, Robertson KD. DNA methylation: superior or subordinate in the epigenetic hierarchy? Genes Cancer. 2011, Jun;2(6):607–17.
24. Baets, J. et al. Defects of mutant DNMT1 are linked to a spectrum of neurological disorders. Brain. 2015, 138, 845–861.
25. Zhang, Z.-M. et al. Structural basis for DNMT3A-mediated de novo DNA methylation. Nature, 2018, 554, 387–391.
26. Yagi, M. et al. Identification of distinct loci for de novo DNA methylation by DNMT3A and DNMT3B during mammalian development. Nat. Commun. 2020,11, 3199.
27. He, Y.-F.; Li, B.-Z.; Li, Z et al. Tet-mediated formation of 5-carboxylcytosine and its excision by TDG in mammalian DNA. Science 2011, 333, 1303–1307.
28. Ito, S.; Shen, L.; Dai, Q.; et al. Tet proteins can convert 5-methylcytosine to 5-formylcytosine and 5-carboxylcytosine. Science. 2011, 333, 1300–1303.
29. Verma, N. et al. TET proteins safeguard bivalent promoters from de novo methylation in human embryonic stem cells. Nat. Genet. 2018, 50, 83–95.
30. Bick, A. G. et al. Genetic interleukin 6 signaling deficiency attenuates cardiovascular risk in clonal hematopoiesis. Circulation. 2020, 141, 124–131.
31. Ahuja, N. et al. Aging and DNA methylation in colorectal mucosa and cancer. Cancer Res. 1998, 58, 5489–5494.
32. Talens, R. P. et al. Epigenetic variation during the adult lifespan: cross-sectional and longitudinal data on monozygotic twin pairs. Aging Cell. 2012, 11, 694–703.
33. Lowe, R. et al. Ageing-associated DNA methylation dynamics are a molecular readout of lifespan variation among mammalian species. Genome Biol. 2018, 19, 22.
34. Horvath, S. DNA methylation age of human tissues and cell types. Genome Biol. 2013, 14, R115.
35. Horvath, S. Erratum to: DNA methylation age of human tissues and cell types. Genome Biol. 2015, 16, 96.
36. Hannum G., Guinney J., Zhao L., et al. Genome-wide methylation profiles reveal quantitative views of human aging rates. *Mol. Cell.* 2013; 49: 359–367

37. McClay, J. L., Aberg, K. A., Clark, S. L., et al. A methylome-wide study of aging using massively parallel sequencing of the methyl-CpG-enriched genomic fraction from blood in over 700 subjects. Hum. Mol. Genet. 2014, 23, 1175–1185.
38. Sidler, C., Woycicki, R., Kovalchuk, I., et al. WI-38 senescence is associated with global and site-specific hypomethylation. Aging. 2014, 6, 564–574.
39. Kouzarides, T. Chromatin modifications and their function. Cell. 2007, 128, 693–705.
40. Jenuwein T. Re-SET-ting heterochromatin by histone methyltransferases. Trends Cell Biol. 2001, Jun;11(6):266–73.
41. Byvoet P, Shepherd GR, Hardin JM, et al. The distribution and turnover of labeled methyl groups in histone fractions of cultured mammalian cells. Arch Biochem Biophys. 1972; 148:558– 67.
42. Fischle W, Franz H, Jacobs SA, et al. Specificity of the chromodomain Y chromosome family of chromodomains for lysine-methylated ARK(S/T) motifs. J Biol Chem. 2008, 283:19626–35.
43. Murray K. The Occurrence of Epsilon-N-Methyl Lysine in Histones. Biochemistry. 1964, 3:10–5.
44. Borun TW, Pearson D, Paik WK. Studies of histone methylation during the HeLa S-3 cell cycle. J Biol Chem. 1972, 247:4288–98.
45. Gershey EL, Haslett GW, Vidali G, et al. Chemical studies of histone methylation. Evidence for the occurrence of 3-methylhistidine in avian erythrocyte histone fractions. J Biol Chem. 1969, 244:4871–7.
46. Greer EL, Shi Y. Histone methylation: a dynamic mark in health, disease and inheritance. Nat Rev Genet. 2014, 13(5): 343–357.
47. Paik WK, Kim S. E-N-dimethyllysine in histones. Biochemical and biophysical research communications. 1967, 27:479–83.
48. Hempel K, Lange HW, Birkofer L. Epsilon-N-trimethyllysine, a new amino acid in histones. Die Naturwissenschaften. 1968, 55:37.
49. Sen, P.; Dang, W.; Donahue, G.; et al. H3K36 methylation promotes longevity by enhancing transcriptional fidelity. Genes Dev. 2015, 29, 1362–1376.
50. Pu, M.; Ni, Z.; Wang, M.; et al. Trimethylation of Lys36 on H3 restricts gene expression change during aging and impacts life span. Genes Dev. 2015, 29, 718–731.
51. Su et al., 2018, Muscle-Specific Histone H3K36 Dimethyltransferase SET-18 Shortens Lifespan of *Caenorharbditis elegans* by Repressing daf-16a Expression. Cell Reports. 2008, March 06(22), 2716–2729.
52. Hertweck, M., Gobel, C., and Baumeister, R. *C. elegans* SGK-1 is the critical component in the Akt/PKB kinase complex to control stress response and life span. Dev. Cell. 2004, 6, 577–588.
53. Paradis S, Ruvkun G. *Caenorhabditis elegans* Akt/PKB transduces insulin receptor-like signals from AGE-1 PI3 kinase to the DAF-16 transcription factor. Genes Dev. 1998, Aug 15;12(16):2488–98.
54. Paradis S, Ailion M, Toker A, Thomas JH, Ruvkun G. A PDK1 homolog is necessary and sufficient to transduce AGE-1 PI3 kinase signals that regulate diapause in *Caenorhabditis elegans*. Genes Dev. 1999, Jun 1;13(11):1438–52.
55. Hsu, A.L., Murphy, C.T., and Kenyon, C. Regulation of aging and age-related disease by DAF-16 and heat-shock factor. Science. 300, 1142–1145.
56. Lin, K., Dorman, J.B., Rodan, A., et al. Daf-16: An HNF-3/ forkhead family member that can function to double the life-span of *Caenorhabditis elegans*. Science. 1997, 278, 1319–1322.
57. Murphy, C.T., McCarroll, S.A., Bargmann, CI., et al. Genes that act downstream of DAF16 to influence the lifespan of *Caenorhabditis elegans*. Nature. 2003, 424, 277–283.
58. Ogg, S., Paradis, S., Gottlieb, S., et al. The Fork head transcription factor DAF-16 transduces insulin-like metabolic and longevity signals in *C. elegans*. Nature. 1997, 389, 994–999
59. Siebold, A.P.; Banerjee, R.; Tie, F.; et al. Polycomb Repressive Complex 2 and Trithorax modulate Drosophila longevity and stress resistance. Proc. Natl. Acad. Sci. USA. 2010, 107, 169–174.

60. Chang KT, Min KT. Regulation of lifespan by histone deacetylase. Ageing Res Rev. 2002, Jun;1(3):313–26.
61. Peleg S, Feller C, Ladurner AG, et al. The Metabolic Impact on Histone Acetylation and Transcription in Ageing. Trends Biochem Sci. 2016, 41, 700–711.
62. Sen, P. et al. Histone acetyltransferase p300 induces de novo super-enhancers to drive cellular senescence. Mol. Cell. 2019, 73, 684–698 e688.
63. Li TY, Sleiman MB, Li H, Gao AW, et al. The transcriptional coactivator CBP/p300 is an evolutionarily conserved node that promotes longevity in response to mitochondrial stress. Nat Aging. 2021, Feb;1(2):165–178.
64. Barrett, L. N. & Westerheide, S. D. The CBP-1/p300 Lysine Acetyltransferase Regulates the Heat Shock Response in *C. elegans*. Front. Aging. 2022, 3, 861761.
65. Wang, W. et al. A genome-wide CRISPR-based screen identifies KAT7 as a driver of cellular senescence. Sci. Transl. Med. 2021, 13, eabd2655.
66. Tasselli, L. et al. SIRT6 deacetylates H3K18ac at pericentric chromatin to prevent mitotic errors and cellular senescence. Nat. Struct. Mol. Biol. 2016, 23, 434–440.
67. Feser, J. et al. Elevated histone expression promotes life span extension. Mol. Cell. 2010, 39, 724–735.
68. Dang, W. et al. Histone H4 lysine 16 acetylation regulates cellular lifespan. Nature. 2009, 459, 802–807.
69. Hachinohe, M., Hanaoka, F. & Masumoto, H. Hst3 and Hst4 histone deacetylases regulate replicative lifespan by preventing genome instability in Saccharomyces cerevisiae. Genes Cells. 2011, 16, 467–477.
70. Brachmann CB, Sherman J M, Devine S E, et al. The SIR2 gene family, conserved from bacteria to humans, functions in silencing, cell cycle progression, and chromosome stability. *Genes & Dev.* 1995, 9: 2888–2902. https://doi.org/10.1101/gad.9.23.2888
71. Guarente, L., Sir2 links chromatin silencing, metabolism, and aging. Genes Dev. 2000, 14, 1021–1026
72. Imai SI, Armstrong CM, Kaeberlein M, et al. Transcriptional silencing and longevity protein Sir2 is an NAD-dependent histone deacetylase. *Nature.* 2000, volume 403, pages795–800.
73. Michishita, E.; Park, J.K.; Burneskis, J.M.; et al. Evolutionarily conserved and nonconserved cellular localizations and functions of human SIRT proteins. Mol. Biol. Cell. 2005, 16, 4623–4635.
74. Watroba, M.; Dudek, I.; Skoda, M.; et al. Sirtuins, epigenetics and longevity. Ageing Res. Rev. 2017, 40, 11–19.
75. Pillai, V.B.; Bindu, S.; Sharp, W.; et al. SIRT3 protects mitochondrial DNA damage and blocks the development of doxorubicin-induced cardiomyopathy in mice. Am. J. Physiol. Heart Circ. Physiol. 2016, 310, H962–H972.
76. Jablonski, R.P.; Kim, S.-J.; Cheresh, P.; et al. SIRT3 deficiency promotes lung fibrosis by augmenting alveolar epithelial cell mitochondrial DNA damage and apoptosis. FASEB J. 2017, 31, 2520–2532
77. Tanno, M., Sakamoto, J., Miura, T., et al. Nucleocytoplasmic shuttling of the NAD+-dependent histone deacetylase SIRT1. J. Biol. Chem. 2007, 282 (9), 6823–6832.
78. Sundaresan, N.R., Samant, S.A., Pillai, V.B., et al. SIRT3 is a stress-responsive deacetylase in cardiomyocytes that protects cells from stress-mediated cell death by deacetylation of Ku70. Mol. Cell. Biol.2008, 28 (20), 6384–6401.
79. Iwahara, T., Bonasio, R., Narendra, V., et al. SIRT3 functions in the nucleus in the control of stress-related gene expression. Mol. Cell. Biol. 2012, 32 (24), 5022–5034.
80. Wilking, M.; Singh, C.K.; Nihal, M.; et al. Sirtuin deacetylases: A new target for melanoma management. Cell Cycle 2014, 13, 2821–2826.
81. Liu, T.; Liu, P.Y.; Marshall, G.M. The critical role of the class II histone deacetylase SIRT1 in cancer. Cancer Res. 2009, 69, 1702–1705.
82. Ong, A.L.C.; Ramasamy, T.S. Role of Sirtuin1-p53 regulatory axis in aging, cancer and cellular reprogramming. Ageing Res. Rev. 2018, 43, 64–80.

83. XING X, ZHANG J, WU T, et al. SIRT1 reduces epigenetic and non-epigenetic changes to maintain the quality of postovulatory aged oocytes in mice. Exper. Cell Res. 2021, 399 112421.
84. Gong, H. et al. Age-dependent tissue expression patterns of Sirt1 in senescence accelerated mice. Mol. Med. Rep. 2014, 10, 3296–3302.
85. Cho, S. H. et al. SIRT1 deficiency in microglia contributes to cognitive decline in aging and neurodegeneration via epigenetic regulation of IL-1beta. J. Neurosci. 2015, 35, 807–818.
86. Erichsen L, Adjaye J. Crosstalk between age accumulated DNA-damage and the SIRT1-AKT-GSK3ß axis in urine derived renal progenitor cells. Aging (Albany NY). 2022, Sep 24;14(20):8179–8204.
87. Ions LJ, Wakeling LA, Bosomworth HJ, et al. Effects of Sirt1 on DNA methylation and expression of genes affected by dietary restriction. Age. 2013, 35:1835–49.
88. Guarente L, Picard F. Calorie restriction—the SIR2 connection. Cell. 2005, 120(4):473–482.
89. Wakeling LA, Ions LJ, Escolme SM, et al. SIRT1 affects DNA methylation of Polycomb group protein target genes, a hotspot of the epigenetic shift observed in ageing. Human Genomics. 2015, 9:14.
90. Hamilton, B. et al. A systematic RNAi screen for longevity genes in *C. elegans*. *Genes Dev.* 2005, 19, 1544–1555.
91. Li, J. et al. *Caenorhabditis elegans* HCF-1 functions in longevity maintenance as a DAF-16 regulator. *PLoS Biol.* 2008, 6, e233.
92. McColl, G. et al. Pharmacogenetic analysis of lithium-induced delayed aging in *Caenorhabditis elegans*. *J. Biol. Chem.* 2008, 283, 350–357.
93. Chen, S. et al. The conserved NAD(H)-dependent corepressor CTBP-1 regulates *Caenorhabditis elegans* life span. *Proc. Natl Acad. Sci. USA*. 2009, 106, 1496–1501.
94. Greer, E., Maures, T., Hauswirth, A. *et al.* Members of the H3K4 trimethylation complex regulate lifespan in a germline-dependent manner in *C. elegans*. *Nature.* 2010, 466, 383–387.
95. Li, L.; Greer, C.; Eisenman, R.N.; et al. Essential functions of the histone demethylase lid. PLoS Genet. 2010, 6, e1001221.
96. Pu, M.; Wang, M.; Wang, W.; et al. Unique patterns of trimethylation of histone H3 lysine 4 are prone to changes during aging in *Caenorhabditis elegans* somatic cells. PLoS Genet. 2018, 14, e1007466.
97. Cruz, C.; Della Rosa, M.; Krueger, C.; et al. Tri-methylation of histone H3 lysine 4 facilitates gene expression in ageing cells. Elife 2018, 7.
98. Li C-L, Pu M, Wang W, et al. Region-specific H3K9me3 gain in aged somatic tissues in *Caenorhabditis elegans*. PLoS Genet. 2021, 17(9): e1009432.
99. Lorbeck, M.T.; Singh, N.; Zervos, A.; et al. The histone demethylase Dmel\Kdm4A controls genes required for life span and male-specific sex determination in Drosophila. Gene 2010, 450, 8–17.
100. Djeghloul, D.; Kuranda, K.; Kuzniak, I.; et al. Age-Associated Decrease of the Histone Methyltransferase SUV39H1 in HSC Perturbs Heterochromatin and B Lymphoid Differentiation. Stem Cell Reports. 2016, 6, 970–984.
101. Vaquero, A.; Scher, M.; Erdjument-Bromage, H.; et al. SIRT1 regulates the histone methyltransferase SUV39H1 during heterochromatin formation. Nature. 2007, 450, 440–444.
102. Sarg, B.; Koutzamani, E.; Helliger, W.; et al. Postsynthetic trimethylation of histone H4 at lysine 20 in mammalian tissues is associated with aging. J. Biol. Chem. 2002, 277, 39195–39201. [CrossRef]
103. Lyu G, Guan Y, Zhang C, et al. TGF-β signaling alters H4K20me3 status via miR-29 and contributes to cellular senescence and cardiac aging. Nat Commun. 2018, Jul 2;9(1):2560.
104. Kudithipudi S, Schuhmacher MK, Kebede AF, et al. The SUV39H1 Protein Lysine Methyltransferase Methylates Chromatin Proteins Involved in Heterochromatin Formation and VDJ Recombination. *ACS Chemical Biology.* **2017,** *12* (4), 958–968. https://doi.org/10.1021/acschembio.6b01076
105. Vaquero A, Scher M, Lee D, et al. Human SirT1 interacts with histone H1 and promotes formation of facultative heterochromatin. Mol Cell. 2004, Oct 8;16(1):93–105. https://doi.org/10.1016/j.molcel.2004.08.031.

106. Wells MB, Csankovszki G, Custer LM. Finding a balance: how diverse dosage compensation strategies modify histone h4 to regulate transcription. *Genet Res Int.* 2012, 2012:795069 Oct 19. https://doi.org/10.1155/2012/795069.
107. Vaquero A, Sternglanz R, Reinberg D. NAD+-dependent deacetylation of H4 lysine 16 by class III HDACs. Oncogene. 2007. 13;26(37):5505–20. https://doi.org/10.1038/sj.onc.1210617.
108. Ma P, Schultz RM. Histone deacetylase 2 (HDAC2) regulates chromosome segregation and kinetochore function via H4K16 deacetylation during oocyte maturation in mouse. PLoS Genet. 2013;9(3):e1003377. https://doi.org/10.1371/journal.pgen.1003377.
109. Kawakami, K.; Nakamura, A.; Ishigami, A.; et al. Age-related difference of site-specific histone modifications in rat liver. Biogerontology 2009, 10, 415–421.
110. Michishita, E.; McCord, R.A.; Berber, E.; et al. SIRT6 is a histone H3 lysine 9 deacetylase that modulates telomeric chromatin. Nature 2008, 452, 492–496.
111. Kaeberlein, M., McVey, M., Guarente, L., The SIR2/3/4 complex and SIR2 alone promote longevity in *Saccharomyces cerevisiae* by two different mechanisms. Genes Dev.1999 13, 2570–2580.
112. Tissenbaum, H.A., Guarente, L., 2001. Increased dosage of a sir-2 gene extends lifespan in *Caenorhabditis elegans*. Nature. 2001, 410, 227–230.
113. Rogina, B., Helfand, S.L. Sir2 mediates longevity in the fly through a pathway related to calorie restriction. Proc. Natl. Acad. 2004, 101, 15998–16003. https://doi.org/10.1073/pnas.0404184101.
114. Bitterman, K.J., Anderson, R.M., Cohen, H.Y., et al. Inhibition of silencing and accelerated aging by nicotinamide, a putative negative regulator of yeast sir2 and human SIRT1. J. Biol. Chem. 2002, 277, 45099–45107. https://doi.org/10.1074/jbc.M205670200.
115. Satoh, A., Brace, C.S., Rensing, N., et al. SIRT1 extends life span and delays aging in mice through the regulation of Nk2 homeobox 1 in the DMH and LH. Cell Metab. 2013, 18, 416–430. https://doi.org/10.1016/j.cmet.2013.07.013.
116. Kauppinen A, Suuronen T, Ojala J, et al. Antagonistic crosstalk between NF-kB and SIRT1 in the regulation of inflammation and metabolic disorders. 2013, Cellular Signalling. 25: 1939–1948.
117. Chen C, Zhou M, Ge Y, et al. SIRT1 and aging related siganling pathways. Mechanisms of Aging and Development. 2020, 187. 111215.
118. Yeung, F., Hoberg, J.E., Ramsey, C.S., et al. Modulation of NF-kappa B- dependent transcription and cell survival by the SIRT1 deacetylase. EMBO J. 2004, 23 (12), 2369–2380.
119. Yang XD, Tajkhorshid E, Chen LF. Functional interplay between acetylation and methylation of the RelA subunit of NF-kappaB. Mol Cell Biol. 2010, May;30(9):2170–80. https://doi.org/10.1128/MCB.01343-09.
120. Ruderman NB, Xu XJ, Nelson L, *et al.* AMPK and SIRT1: a long-standing partnership? *Am J Physiol Endocrinol Metab.* 2010 **298**: E751–E760.
121. Sung, B., Chung, J.W., Bae, H.R., et al. Humulus japonicus extract exhibits antioxidative and anti-aging effects via modulation of the AMPK-SIRT1 pathway. Exp. Ther. Med. 2015, 9 (5), 1819–1826. https://doi.org/10.3892/etm.2015.2302.
122. Lan F, Cacicedo JM, Ruderman N, et al. SIRT1 modulation of the acetylation status, cytosolic localization, and activity of LKB1. Possible role in AMP-activated protein kinase activation. *J Biol Chem.* 2008, 283: 27628–27635.
123. Canto C, Gerhart-Hines Z, Feige JN, *et al.* AMPK regulates energy expenditure by modulating NAD+ metabolism and SIRT1 activity. *Nature.* 2009; 458: 1056–1060.
124. Zhang, Y., Mi, S.L., Hu, N., et al. Mitochondrial aldehyde dehydrogenase 2 accentuates aging-induced cardiac remodeling and contractile dysfunction: role of AMPK, SIRT1, and mitochondrial function. Free Radic. Biol. Med. 2014, 71, 208–220.
125. Kitada M, Ogura Y, Monno I, et al. Sirtuins and type 2 diabetes: role in inflammation, oxidative stress, and mitochondrial function. Front. Endocrinol. (Lausanne). 2019, 10: 187.
126. Dilberger, B., Baumanns, S., Schmitt, F., et al. Mitochondrial oxidative stress impairs energy metabolism and reduces stress resistance and longevity of *C. elegans*. 2019, Oxid. Med. Cell. Longev. 2019, 6840540.

127. Chen, P., Chen, F.C., Lei, J.X., et al. Activation of the miR-34a-mediated SIRT1/mTOR signaling pathway by urolithin A attenuates D- galactose-induced brain aging in mice. Neuro Ther. 2019, 16, 1–14.
128. Ghosh, H.S., McBurney, M., Robbins, P.D. SIRT1 negatively regulates the mammalian target of rapamycin. PLoS One. 2010, e9199. https://doi.org/10.1371/journal.pone.0009199.
129. Ou, X., Lee, M.R., Huang, X., et al. SIRT1 positively regulates autophagy and mitochondria function in embryonic stem cells under oxidative stress. Stem Cells. 2014, 32, 1183–1194.
130. Brunet, A., Sweeney, L.B., Sturgill, J.F., et al. Stress-dependent regulation of FOXO transcription factors by the SIRT1 deacetylase. Science. 2004 303 (5666), 2011–2015.
131. Langley E, Pearson M, M. Faretta M, et al. Human SIR2 deacetylates p53 and antagonizes PML/p53-induced cellular senescence. EMBO J. 2002, 21 (10):2383–2396.
132. Varizi H, Dessain SK, Eaton E et al. Hsir2 (SIRT1) functions as an NAD-dependent p53 deacetylase. Cell. 2001, 107(2):159–159.
133. Glozak MA, Sengupta N, Zhang X, et al. Acetylation and deacetylation of non-histone proteins. Gene. 2005, Dec 19;363:15–23. https://doi.org/10.1016/j.gene.2005.09.010.

Chapter 5
Proteostasis, Proteodynamics and Cellular Aging

Jacek Maciej Witkowski, Ewa Bryl, and Tamas Fulop

Introduction

Although the calculated number of proteins (in fact peptides) which are the result of translation from protein-coding genes in human genome only slightly exceeds 20 thousand, the real numbers of different proteins interacting in the cells are by orders of magnitude larger [1]. It is so because of many metabolic events that interfere and result in manufacturing of final, mature and functional proteins. These include first of all combination of "just translated" peptides into a quaternary structure of at least two such peptides, identical or not. This effect is achieved for some proteins by creation of disulfide bonds between the relevant amino-acids of both peptides; such bonds are at least semi-permanent. Then, each nascent peptide may be modified in many ways, including limited proteolytic cleavage, where at least a part of it is removed as unnecessary for the final protein, or the original peptide is cleaved into smaller, yet functional parts. The obvious example here is the proopiomelanocortin (POMC) which is translated as a single precursor peptide of 267 amino-acids, but then post-translationally cleaved to first eliminate the 26 amino-acid terminal of so called signal peptide. Then, remaining 241 amino-acid-long peptide is further cleaved resulting in the formation of α-MSH, ACTH and endogenous opiates (-βendorphin and [met]enkephalin) [2]. Thus, out of a single translation product we end up with four different, functional proteinaceous entities, and this is not a unique, single case.

J. M. Witkowski (✉)
Department of Embryology, Medical University of Gdańsk, Gdańsk, Poland
e-mail: jacek.witkowski@gumed.edu.pl

E. Bryl
Department of Pathophysiology, Medical University of Gdańsk, Gdańsk, Poland

T. Fulop
Geriatric Division, Department of Medicine, Faculty of Medicine and Health Sciences, Research Center on Aging, University of Sherbrooke, Sherbrooke, QC, Canada

V. Bueno (ed.), *Cellular and Molecular Aspects of Ageing*,
https://doi.org/10.1007/978-3-031-55022-5_5

And this is not all what can happen to a nascent peptide on its path to becoming a mature protein.

Then, there is a huge set of other posttranslational modification events [3]. These include first of all glycosylation (where one peptide may be equipped with different oligosaccharide chains, making glycoproteins with different functionalities—essentially different proteins). It was calculated that the number of different glycoproteins (having not only different properties, but first of all different functions in the cells or extracellular spaces) reaches the staggering 10 million (!) species [1].

Next popular posttranslational modification of nascent peptides is their lipidation (addition of lipids including prenylation (especially farnesylation), N-terminal palmitoylation of cysteine residues, and addition of fatty acid residues to serines and lysines of such peptide. Importantly, lipidation makes such modified proteins capable of associating with cell membranes, both outside (serving e.g. as some receptors) or in organelle membranes.

Relatively simple (considering chemistry), but also very important from the point of view of changed functionality, would be hydroxylation, acetylation or methylation of relevant amino-acid residues, affecting the abilities of a given protein to interact with other ones or just rendering it (temporarily or permanently) (in)active. In this category, the most prominent modification would be protein phosphorylation (by a variety of protein kinases) and dephosphorylation by dedicated phosphatases; this serves as an on–off switch for relevant protein activity.

A special and important form of posttranslational modification of peptides and proteins would be their SUMOylation [4]. SUMOs are the small, ubiquitin-like modifier peptides which attach to the relevant, modified proteins enabling different functionalities, including nuclear-cytosolic transport, transcriptional regulation, affecting the proteins' stability and response to stress, and regulating progression through the cell cycle and finally apoptosis of the cells.

And finally ubiquitination; the modification of protein which eventually terminates its "life", as the ubiquitinated proteins are targeted to be degraded in proteasomes [5]. This posttranslational modification belongs to the strictly defined proteostasis.

All these posttranslational modifications other than glycosylation are responsible for formation of ± a hundred thousand different protein species [1]. Of course, the repertoire of different protein species in any given cell type (human or animal) is characteristic and unique for this specific cell type and forms their proteome. All of these different proteins perform their relevant cellular (or extracellular) functions, interacting with other (proteinaceous or not) cellular components. Totality of these interactions forms the interactome of the specific cell type.

Proteostasis or Proteodynamics?

One needs to consider that the quantity, quality, characteristics and functionality of any mature protein depends not only on the process of translation and all the posttranslational modifications mentioned above. At the beginning there must be a peptide-coding gene, which is then transcribed to mRNA, which is then (to make long story short) translated to the unique peptide. Such gene must first be triggered for activation by transcription factor(s) being the final result of activation of specific signal transduction pathways, usually started by ligation of some receptor (membrane or cytoplasmic with the relevant ligand (hormone, growth factor, cytokine etc.) The amount of transcript is being tightly regulated and it will affect the amount of translated peptide. One way of such regulation will be the epi-genetic processes involving DNA (methylation/demethylation) and related proteins (mainly histones which undergo acetylation/deacetylation which changes their affinity to DNA and enables transition from heterochromatin to euchromatin and vice versa). Then, the resulting mature, functional proteins are by no means permanent. Their functional half-life time is very variegated; spanning times from minutes (for obvious reasons for some signaling molecules, e.g. insulin) to as many as 100+ years for some species of human extracellular collagen. Even when we consider intracellular proteins only, their half-life may range from minutes to hours to many (100+) days [6, 7]. Then, each and every cellular protein will undergo time-dependent structural and functional deterioration leading first of all to its malfunction and secondarily to the formation of aggregates of such dysfunctional proteins which may interfere in the function of cells as the whole and as such should be eliminated. Here comes the proteostasis in its strict definition, understood as the homeostatic set of processes aiming at detection and elimination of misfolded, aggregated and otherwise changed cellular proteins and thus improving the "health" of a cell [8, 9]. This goal is achieved by directing these proteins to proteasomes (after their abovementioned ubiquitination) or to lysosomes, where they undergo degradation by acidic proteases and peptidases, including the cathepsins. The facilitators of both these processes is the family of chaperones or heat shock proteins, which detect and bind faulty proteins and then direct them to either proteasome or lysosome [8, 9].

Thus, protein turnover seems to be a necessary way of ascertaining that the proteome would work properly and seamlessly. It was postulated that the longer lifespan of a protein the more prone it would be to deterioration and eventual dysfunctionality. Also, a concept of long-living proteins as the ones that may harbor more and more defects with time and so participate in the proteome aging has been introduced [6, 7].

Finally, the well-being or cellular proteome is considered so important for the well-being of the cell as a whole, that proteostasis was included among the hallmarks of cellular aging [10, 11].

Recently, aiming at better understanding of changes in the cellular proteomes associated with aging, we have introduced the concept of proteodynamics as the totality of ALL processes from induction of peptide-coding gene by relevant transcription factors to transcription itself to translation to postranslational modifications creating functional proteins to protein (and the above mentioned processes malfunction)-related quantitative and qualitative deterioration to final proteostatic degradation [12–15]. The effect of these adverse, age-associated processes leads to changes in cellular proteomes and creation of "old" proteomes which we dubbed the "gero-proteomes"; the process of geroproteome formation we call the "proteostenosis" [12–15].

Why the Proteins Deteriorate Quantitatively and Qualitatively with Time/Age?

We need to remember that each and every stage in the process of manufacturing of any functional protein is performed by specialized proteins which may also undergo all age-dependent changes described below. This way the whole concept of defective proteostasis/proteodynamics in aged cells acquires a circular pathomechanism.

Thus, each and every proteodynamic process may be and generally is affected by cellular and organismal aging. In fact it starts with the deterioration of signal transduction pathways starting with interaction between some cellular receptor and its ligand (e.g. the T cell receptor/CD3 complex and the antigenic peptide ligating at the T cell surface at the beginning or an immune reaction) and ending with activation of specific transcription factors that would stimulate transcription of specific genes (e.g. that of interleukin-2 (IL-2) in stimulated T cells) [16, 17]. Then, the coding DNA itself, as well as its cis and trans regulatory elements may be affected by aging (including accumulation of mutations), leading to minute errors in transcribed code, as well as to decreased efficiency of transcription. Another modification of DNA availability for transcription will be changing epigenetic modifications, including *inter* many *alia* methylation of DNA itself and acetylation of DNA-associated histones. It was found that these and other processes defining the epigenome undergo multiple changes with aging of various cell types [18–20]. The result of these changes to the genome and epigenome of aging cells may be fewer mRNA molecules, additionally burdened with single nucleotide changes in the sequence; the later may result in inclusion of "wrong" amino-acid in the translated peptide [18–20].

Transcription process itself may with cellular aging suffer from quantitative issues (fewer mRNA molecules due to decreased effectiveness of relevant enzymatic systems (transcription machinery) or to reduced availability of nucleotides being the building blocks for the mRNA; also, the cellular energy output (especially resulted from oxidative phosphorylation) is reduced in aging cells, while the protein manufacturing at any stage (including transcription) is highly energy-demanding. The latter will be of course true also for further stages of protein manufacturing, maintenance and disposal. Relevant qualitative issues (apart from already mentioned possible errors in the transcribed DNA) may include decreased fidelity of transcription machinery itself, increasing the numbers of different errors in resulting mRNA.

Of course we need to remember that transcription yields not only the mRNAs, but also other types of ribonucleic acid, including ribosomal (rRNA), amino-acid transporting—transfer (tRNA) and multiple species of small, non-coding regulatory RNA like microRNA, sncRNA, lncRNA etc. While the rRNA (being the building blocks of ribosomes), and tRNAs which provide amino-acid residues for the new peptide building based on mRNA matrix directly correspond to resulting peptide quality and quantity, small regulatory RNAs may affect mainly the process of transcription or the fate of mRNA in the time between transcription and the onset of translation; that way they define to the extent the final peptide yield [21]. It was shown that different species of these small regulatory RNAs change in quantity and quality with aging [21].

Recently a class of molecules called RNA-binding proteins (RBPs) has been discovered, studied and implicated not only in regulation of gene expression by determining the post-transcriptional fate of transcripts, including their maturation, trafficking, stabilization, RNA splicing, export, cellular localization, and degradation as well as translation, but also (or by interfering with these processes) emerging as key regulators of cellular aging and senescence [22].

Even if the quality of actual transcript is perfect, and its quantity reflects the need of the cell for the relevant protein, there are many following steps which can go wrong, leading to decreased quality and/or quantity of final protein. Thus, the mechanism of translation requires the mRNA to be read and "decoded" in the ribosomes (free or attached to the rough endoplasmic reticulum) where, based on the mRNA code, consecutive amino-acids are introduced by the corresponding tRNAs, induced to form peptide bonds by respective initiation factor(s) (IFs), and then continuing to elongate the peptide with the help of elongation factor(s) (EFs). So, for proper and efficient execution of translation the cell needs (besides the mRNA) adequate numbers of working ribosomes, sufficient amino-acid supply, the whole repertoire of tRNAs, an active IFs and EFs (activated by relevant kinases). All of these components undergo aging-associated changes, eventually leading to decrease rates of protein production [23, 24]. On top of these issues there is also the question of translational fidelity, i.e., what is the rate of erroneous building-in a wrong amino-acid in the nascent peptide. It was found that this rate is also growing with cellular age, leading to manufacturing of less functional peptides and sometimes the ones which would not perform as final proteins but rather misfold and/or aggregate [23–25].

The next stage in maturation of nascent proteins is their posttranslational modifications (PTMs). These would include appropriate folding of initially straight linear peptides into primary, secondary, tertiary and quaternary structures. This folding depends on the amino-acid sequences of the folding peptides (i.e., it may be affected by translational errors leading to modified sequence) and then on above mentioned chemical modification of each interacting peptides. It was demonstrated that all forms of chemical modifications of nascent proteins undergo aging-associated changes.

One of the most prominent is the accumulation of non-enzymatically glycosylated (glycated) proteins which lose their functions and tend to aggregate [23–25].

Then, SUMOylation seems to be crucially involved in cellular senescence. The mechanism involves SUMOylation of p53 which then translocates to the PML nuclear bodies where it interacts with other SUMOylated factors, eventually activating the pRb. In turn, cooperation of active pRb with many other SUMOylated proteins creates a repressor complex inhibiting the transcription of proliferation-associated genes and initiating chromatin condensation associated with cellular senescence [4]. Next posttranslational modification of proteins considered important for the process of cellular senescence is their excessive carbonylation, being the direct consequence of oxidative stress and damage which increase with aging [26]. Other PMTs becoming more abundant in aging cells are deamidation, racemization, cross-linking, methylation, carbamylation and truncation. The latter, truncation, is the shortening of nascent peptide due to modification of mRNA-coding DNA in a way cresting an early STOP codon or due to proteolytic cleavage (modifying proteolysis—see below) [27]. Resulting protein is shorter, but in some cases it may retain some (albeit altered) function.

In addition to glycation, apart from affecting proteins in healthy aging cells, the variegated PTMs in aged cells cause inter alia the aging-associated cataracts, by affecting the lens crystallins and other proteins [28]. Also, carbamylation and other PTMs participate in other aging-related conditions, including the chronic kidney disease (CKD)—the multisystem disorder eventually associated with impaired mitochondrial biogenesis and function and low-grade inflammation (inflammaging) [29]. The mechanistic background for these adverse effects of aging-aggravated PTMs may be their modification of some key signaling molecules, including nuclear factor kappa-B (NF-κB), NRF2, and transforming growth factor-β (TGF-β), which undergo ubiquitylation, SUMOylation, phosphorylation, glycosylation, and acetylation. Finally, the posttranslational modifications of proteins which are the most frequent/abundant (90% of all PTMs) and consist of their phosphorylation, acetylation and ubiquitination also undergo changes with cellular aging [3]. It is worth noting here that any PTM may—in the aged cells—undergo increase or decrease, depending on cellular context and possibly other factors; thus, its effects in these aged cells may greatly vary.

Last PTM to be considered here in relation to aging-related modification of proteins is limited modulating proteolysis, where part of a nascent peptide (or, in fact, also of a mature, functional protein) is proteolytically removed in order to render the resulting (shorter) protein active or inactive. In fact most of about 460 of different proteases existing in human cells (and extracellular fluids) perform this extremely important type of PTM, and only a few are the protein degrading ones [27, 30]. Proteases that perform the limited modulating proteolysis include the ACE family, kallikrein, the family of matrix metalloproteases (MMP), renin, cathepsins, plasminogen activator, a family of about 30 ADAM (a disintegrin and metalloprotease) proteases, nineteen ADAMTS (a disintegrin and metalloproteinase with thrombospondin motifs) which participate in extracellular remodeling of collagens and other components of the extracellular matrix and notably the 14-strong family of calpains. Their modulatory activities participate in cellular processes including (but not limited to) cell-cycle progression, replication of DNA, cell proliferation and cell death (notably apoptosis), cellular movements and the immune response. Our own, partly published so far studies concerning changes of calpain amounts and activities in human peripheral blood lymphocytes had shown that aging is associated with significant decrease of both parameters and that successful aging and becoming a longevous individual (centenarian) is associated with retention of amounts and activities of calpains in the PBL [31]. Mechanistically, decreased calpain activities in the lymphocytes of elderly individuals are associated with reduced phosphorylation of key transcription factor NFκB, p56Lck kinase, ZAP70 and phospholipase C gamma, resulting in decreased proliferative and secretory responses of these cells upon activation [32]. It is also possible that these lowered calpain activities in the lymphocytes are at least in part responsible for their increased apoptosis upon stimulation; contrarily we have demonstrated that increased calpain activities in both acute and chronic lymphoid leukemia cells prevent them from undergoing apoptosis, by cleaving the effector caspase-3 [33–35]. Our critical analysis of available publications indicated that this phenomenon (decrease of activities of multiple effectors of modulating proteolysis in the elderly and their retention in centenarians) seems to be general throughout many cell types [15].

Concluding, one can safely assume that different (if not all) forms of PTM are variably modified by cellular and organismal aging, which affects the amounts and properties of proteins being the targets of these PMTs. All these processes from pre-transcription to post-translational modifications belong to proteodynamics or proteostasis in broad sense, and all of them may be affected by aging.

Even the structurally and functionally "ideal" cellular protein would ultimately age, lose its activity and structure which would make it obsolete for the cell. Such proteins, especially modified by glycation or by creation of additional sulfhydryl bonds, tend to misfold or aggregate. As mentioned above, protein (or peptide) misfolding may occur also early in its creation due to translational errors. Whatever the cause, with aging cells tend to accumulate more aggregated and misfolded proteins. This does not happen in young cells due to two main reasons. First, young cells make much less improperly folded proteins or these tending to aggregate. And second, in young cells the unfolded protein response (UPR) is strong and leads to elimination of such intracellular burden by various types of autophagy. This elimination can be called "proteostasis in narrow sense". So, unfolded and aggregated proteins are detected (as mentioned above, with the help of dedicated heat shock proteins (HSP) called the chaperones and targeted to one of the autophagic effectors: lysosomes, proteasomes and cytoplasmatic degrading proteases (like ADAMs). All these different effectors of autophagy (including but not limited to proteostasis sensu stricto) are adversely affected by aging, as all of them are also proteins and undergo the same deteriorating effects of aging described above as other proteins. Thus, the numbers and properties of HSP molecules in aging cells are reduced, making the detection and targeting of misfolded or aggregated proteins to destruction less effective [36]. Then, compared to young cells, lysosomes of aged cells may be fewer, contain less proteolytic enzymes (mainly cathepsins) and their internal acidity (necessary for proper activity of these proteases) dependent on less effective operation of proton pumps in their membranes is also inadequate, so even the remaining lysosomal proteases cannot be fully activated [12, 15]. One has to remember here that proton pumps, similarly to other enzymes shunting whatever ions across the membranes against the concentration gradient require huge amounts of ATP and, as mentioned above, ATP output in old cells is reduced due to mitochondrial aging including affected mitochondrial proteodynamics [37]. Similarly, the activity of proteasomes in aging cells is decreased, due to both decreased numbers of these multiprotein complexes per cell, decreased activities of incorporated proteases, and lower efficiency of ubiquitination targeting the proteins to be degraded in proteasomes due to less effective enzymes and lower available energy [38–40]. With the exception of calpains, cytoplasmic proteases which may be involved in proteolytic modulation of cytoplasmic (and other) proteins seem to be less affected by aging.

Summary and Conclusion

As shown in the above discussion, practically all aspects of proteodynamics in old cells are adversely affected, leading to conversion of vibrant proteomes of young cells to geroproteomes, where multiple proteins are quantitatively and qualitatively deteriorating. This leads to malfunction of various intracellular metabolic pathways which, by interacting, affect other pathways even if these are not yet intrinsically faulty. This creates the vicious circle, eventually leading to the cell senescence and ultimately demise. Still, if we consider healthy, successful aging allowing one to live to 100+ ears of age, it is likely that protein homeostasis is maintained at sufficient state for (much) longer than in average aging and the changes of different aspects of proteodynamics occur slower than in average aging and much slower than in pathological, accelerated aging associated with aging-related diseases, where the proteostenotic changes would be comparatively increased, accelerated, and quickened [12, 15].

From this point of view, proteostasis (sensu *largo*, or proteodynamics) should be not only one of the hallmarks of cellular aging and senescence, but the most important of all of them. Citing the philosopher Frederick Engels, "Life is the mode of existence of protein bodies, the essential element of which consists in *continual metabolic interchange with the natural environment outside them,* and which ceases with the cessation of this metabolism, bringing about the decomposition of the protein" [41].

References

1. Ponomarenko EA, Poverennaya EV, Ilgisonis EV, Pyatnitskiy MA, Kopylov AT, Zgoda VG, et al. The Size of the Human Proteome: The Width and Depth. Int J Anal Chem. 2016;2016:7436849.
2. Pritchard LE, White A. Neuropeptide processing and its impact on melanocortin pathways. Endocrinology. 2007;148(9):4201–7.
3. Ebert T, Tran N, Schurgers L, Stenvinkel P, Shiels PG. Ageing - Oxidative stress, PTMs and disease. Mol Aspects Med. 2022;86:101099.
4. Scurr LL, Haferkamp S, Rizos H. The Role of Sumoylation in Senescence. Adv Exp Med Biol. 2017;963:215–26.
5. Zhong Q, Xiao X, Qiu Y, Xu Z, Chen C, Chong B, et al. Protein posttranslational modifications in health and diseases: Functions, regulatory mechanisms, and therapeutic implications. MedComm (2020). 2023;4(3):e261.
6. Toyama BH, Hetzer MW. Protein homeostasis: live long, won't prosper. Nat Rev Mol Cell Biol. 2013;14(1):55–61.
7. Fornasiero EF, Savas JN. Determining and interpreting protein lifetimes in mammalian tissues. Trends Biochem Sci. 2023;48(2):106–18.
8. Ghosh Chowdhury S, Ray R, Karmakar P. Relating aging and autophagy: a new perspective towards the welfare of human health. EXCLI J. 2023;22:732–48.
9. Hoppe T, Cohen E. Organismal Protein Homeostasis Mechanisms. Genetics. 2020;215(4):889–901.
10. Lopez-Otin C, Blasco MA, Partridge L, Serrano M, Kroemer G. The hallmarks of aging. Cell. 2013;153(6):1194–217.
11. Lopez-Otin C, Blasco MA, Partridge L, Serrano M, Kroemer G. Hallmarks of aging: An expanding universe. Cell. 2023;186(2):243–78.
12. Frankowska N, Lisowska K, Witkowski JM. Proteolysis dysfunction in the process of aging and age-related diseases. Front Aging. 2022;3:927630.
13. Witkowski JM, Bryl E, Fulop T. Proteodynamics and aging of eukaryotic cells. Mech Ageing Dev. 2021;194:111430.
14. Witkowski JM, Mikosik A, Bryl E, Fulop T. Proteodynamics in aging human T cells - The need for its comprehensive study to understand the fine regulation of T lymphocyte functions. Exp Gerontol. 2018;107:161–8.
15. Frankowska N, Bryl E, Fulop T, Witkowski JM. Longevity, Centenarians and Modified Cellular Proteodynamics. Int J Mol Sci. 2023;24(3).
16. Fulop T, Witkowski JM, Le Page A, Fortin C, Pawelec G, Larbi A. Intracellular signalling pathways: targets to reverse immunosenescence. Clin Exp Immunol. 2017;187(1):35–43.
17. Le Page A, Dupuis G, Larbi A, Witkowski JM, Fulop T. Signal transduction changes in CD4(+) and CD8(+) T cell subpopulations with aging. Exp Gerontol. 2018;105:128–39.
18. Wu Z, Qu J, Zhang W, Liu GH. Stress, epigenetics, and aging: Unraveling the intricate crosstalk. Mol Cell. 2023.
19. Wang K, Liu H, Hu Q, Wang L, Liu J, Zheng Z, et al. Epigenetic regulation of aging: implications for interventions of aging and diseases. Signal Transduct Target Ther. 2022;7(1):374.
20. Klutstein M, Gonen N. Epigenetic aging of mammalian gametes. Mol Reprod Dev. 2023.
21. Caravia XM, Lopez-Otin C. Regulatory Roles of miRNAs in Aging. Adv Exp Med Biol. 2015;887:213–30.
22. Varesi A, Campagnoli LIM, Barbieri A, Rossi L, Ricevuti G, Esposito C, et al. RNA binding proteins in senescence: A potential common linker for age-related diseases? Ageing Res Rev. 2023;88:101958.
23. Woodward K, Shirokikh NE. Translational control in cell ageing: an update. Biochem Soc Trans. 2021;49(6):2853–69.
24. Jaisson S, Desmons A, Gorisse L, Gillery P. [Protein molecular aging: which role in physiopathology?]. Med Sci (Paris). 2017;33(2):176–82.

25. Basha S, Mukunda DC, Rodrigues J, Gail D'Souza M, Gangadharan G, Pai AR, et al. A comprehensive review of protein misfolding disorders, underlying mechanism, clinical diagnosis, and therapeutic strategies. Ageing Res Rev. 2023;90:102017.
26. Baraibar MA, Ladouce R, Friguet B. Proteomic quantification and identification of carbonylated proteins upon oxidative stress and during cellular aging. J Proteomics. 2013;92:63–70.
27. Fortelny N, Pavlidis P, Overall CM. The path of no return--Truncated protein N-termini and current ignorance of their genesis. Proteomics. 2015;15(14):2547–52.
28. Fan X, Monnier VM. Protein posttranslational modification (PTM) by glycation: Role in lens aging and age-related cataractogenesis. Exp Eye Res. 2021;210:108705.
29. Nicolas C, Jaisson S, Gorisse L, Tessier FJ, Niquet-Leridon C, Jacolot P, et al. Carbamylation is a competitor of glycation for protein modification in vivo. Diabetes Metab. 2018;44(2):160–7.
30. Turk B. Targeting proteases: successes, failures and future prospects. Nat Rev Drug Discov. 2006;5(9):785–99.
31. Mikosik A, Foerster J, Jasiulewicz A, Frackowiak J, Colonna-Romano G, Bulati M, et al. Expression of calpain-calpastatin system (CCS) member proteins in human lymphocytes of young and elderly individuals; pilot baseline data for the CALPACENT project. Immun Ageing. 2013;10(1):27.
32. Mikosik A, Jasiulewicz A, Daca A, Henc I, Frackowiak JE, Ruckemann-Dziurdzinska K, et al. Roles of calpain-calpastatin system (CCS) in human T cell activation. Oncotarget. 2016;7(47):76479–95.
33. Mikosik A, Henc I, Ruckemann-Dziurdzinska K, Frackowiak JE, Ploszynska A, Balcerska A, et al. Increased mu-Calpain Activity in Blasts of Common B-Precursor Childhood Acute Lymphoblastic Leukemia Correlates with Their Lower Susceptibility to Apoptosis. PLoS One. 2015;10(8):e0136615.
34. Lopatniuk P, Puchalska Z, Mital A, Mikosik A, Frackowiak J, Hellmann A, et al. Excessive amount and activity of mu-calpain affects apoptotic machinery in chronic B-cell leukemia cells and influences the course of the disease. Acta Biochim Pol. 2020;67(2):247–57.
35. Witkowski JM, Zmuda-Trzebiatowska E, Swiercz JM, Cichorek M, Ciepluch H, Lewandowski K, et al. Modulation of the activity of calcium-activated neutral proteases (calpains) in chronic lymphocytic leukemia (B-CLL) cells. Blood. 2002;100(5):1802–9.
36. Hebishy M, Shintouo CM, Dufait I, Debacq-Chainiaux F, Bautmans I, Njemini R. Heat shock proteins and cellular senescence in humans: A systematic review. Arch Gerontol Geriatr. 2023;113:105057.
37. Chen Q, Young L, Barsotti R. Mitochondria in cell senescence: A Friend or Foe? Adv Protein Chem Struct Biol. 2023;136:35–91.
38. Ponnappan S, Ovaa H, Ponnappan U. Lower expression of catalytic and structural subunits of the proteasome contributes to decreased proteolysis in peripheral blood T lymphocytes during aging. Int J Biochem Cell Biol. 2007;39(4):799–809.
39. Ponnappan S, Uken-Trebilcock G, Lindquist M, Ponnappan U. Tyrosine phosphorylation-dependent activation of NFkappaB is compromised in T cells from the elderly. Exp Gerontol. 2004;39(4):559–66.
40. Ponnappan U. Ubiquitin-proteasome pathway is compromised in CD45RO+ and CD45RA+ T lymphocyte subsets during aging. Exp Gerontol. 2002;37(2–3):359–67.
41. Engels F. Dialectics of Nature Dialectics of Nature. 6 ed: Progress Publishers; 1883. p. 295–311.

Chapter 6
Mitochondria and Ageing

Tiago Rodrigues

Introduction

It is indisputable that the discovery of antibiotics revolutionized the Medicine in the early 1900s, decreasing the number of deaths by microbial infections [17]. After that, advances in therapeutics and medical care, combined with the improvement of basic sanitation and living conditions have significantly increased health and life expectancy. It is clear that population around the world are living longer and most countries are reporting the ageing of population. Therefore, aging and its consequences have drawn the attention of scientists, trying to unveiling aging process and to mitigate possible harmful effects and aging-related diseases.

So, why do we age? This question seems to have an ordinary explanation, such as "it is an engine whose components start to wear out over time". Nevertheless, the answer is not so simple because aging has risen as a major risk factor for several diseases. Actually, scientists worldwide have trying to answer a second question in order to find responses to the first one, which is "how do we age?", i.e., what are the genetics and molecular alterations involved in this complex and progressive biological process. The current knowledge is that aging is unstoppable, but which evolves with different rates according to the person and this raises many other questions. How is this process controlled? Can we deal with it? Can we pharmacologically or genetically manage it? All of these are still open questions, but in this chapter, the roles played by mitochondria in aging will be reviewed and critically discussed, focusing on the molecular aspects and alterations involved in healthy and aging-related diseases.

T. Rodrigues (✉)
Center for Natural and Human Sciences (CCNH), Federal University of ABC (UFABC), Avenida Dos Estados, 5001. Bloco A, Torre 3, Sala 623, Santo Andre, SP CEP 09210-580, Brazil
e-mail: tiago.rodrigues@ufabc.edu.br

V. Bueno (ed.), *Cellular and Molecular Aspects of Ageing*,
https://doi.org/10.1007/978-3-031-55022-5_6

Mitochondria: A Central Hub of Energy and Metabolism

To understand why mitochondria are at the center of discussions about aging, the fundamentals of mitochondrial morphology and functions must be considered. Mitochondria are organelles delimited by a double membrane system comprised by the inner (IMM) and the outer mitochondrial membrane (OMM). These membranes are separated by the intermembrane space. The IMM has several folds that result in invaginations (mitochondrial cristae), and the inner compartment delimited by the IMM is the mitochondrial matrix (Fig. 6.1). Although frequently described as oval-shaped organelles, mitochondria alter their morphology constantly, according to the external stimuli or cellular requirements. Such morphological alterations, named mitochondrial dynamics, are strictly regulated and mediated by proteins that promote fusion and fission, remodeling the mitochondrial network [33, 64]. The OMM are relatively permeable to the cytosol content due to the presence of the voltage-dependent anion channel (VDAC), a b-barrel porin [78]. In contrast, the impermeability of the IMM, including to H^+ (whose importance will be further highlighted), drives the need for a plethora of carrier proteins to mediate the transport of small molecules, solutes, metabolic intermediates, and ions to the matrix [20]. Such compartmentalization is pivotal for the central role of mitochondria in the cellular energy metabolism.

Mitochondria have been considered as the "powerhouse" of cells by producing the major part of the energy required for the cellular maintenance. And this statement is really true! However, during the last decades, accumulating evidence have pointed to the participation of mitochondria in several other important biological processes, including the regulation of Ca^{2+} homeostasis, ROS production, signaling, communication with other organelles and cells, and cell fate (death/survival) [16, 43]. All of these processes enable mitochondria as a central hub in the cell, integrating metabolism and redox state in physiological and pathological conditions, including aging.

Mitochondria are widely recognized to play a central role in cellular metabolism, since they house several metabolic pathways. Acetyl-CoA formed from the oxidative decarboxylation of pyruvate catalyzed by the pyruvate dehydrogenase complex (PDH) or by fatty acid oxidation (b-oxidation) in the mitochondrial matrix is condensed with oxalacetate (OAA) to produce citrate, feeding the tricarboxylic acid cycle (TCA), also known as Krebs' or citric acid cycle. The energy released during the sequential reactions of this cyclic metabolic pathway is conserved as nicotinamide adenine dinucleotide (NADH), flavine adenine nucleotide ($FADH_2$), and guanosine triphosphate (GTP). The reduced coenzymes NADH and $FADH_2$ donate their reducing equivalents to the electron transport chain (ETC). ETC is formed by the respiratory complexes I, II, III, and IV. The entrance of electrons from NADH occurs via complex I and from $FADH_2$ via complex II, which are transferred to complex III by the mobile carrier ubiquinone or coenzyme Q. Electrons are transferred from complex III to complex IV by the other mobile carrier cytochrome *c*. Then, O_2 receive the electrons from redox centers in complex IV to be fully reduced to H_2O.

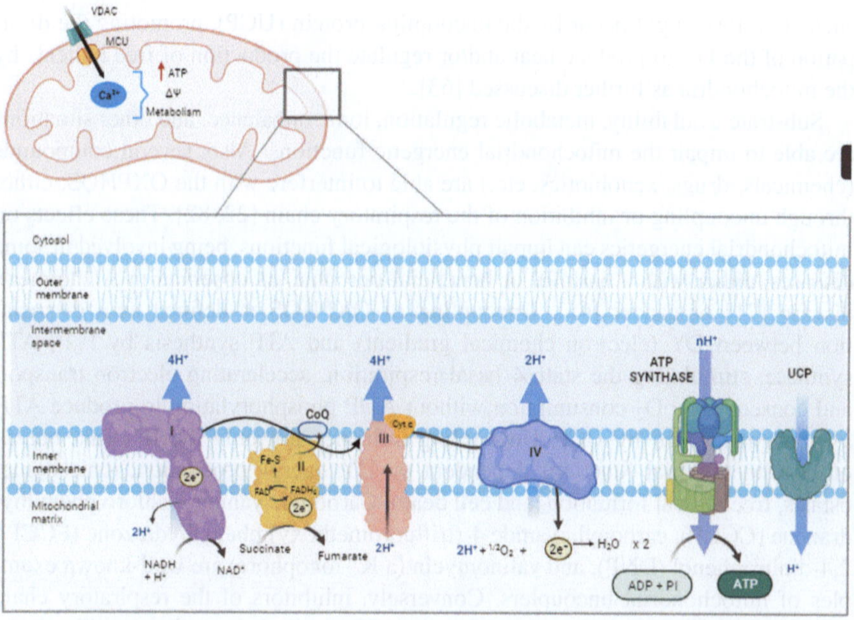

Fig. 6.1 Representative scheme of mitochondrial oxidative phosphorylation, showing the electron transport coupled to the ATP synthesis. Mitochondrial morphology is illustrated in the upper part, highlighting the double membrane system formed by the outer and inner mitochondrial membrane (OMM and IMM), delimiting the cytosolic, intermembrane space, and mitochondrial matrix compartments. The voltage-dependent anion channel (VDAC) allowing the movement of small solutes and ions are presented in OMM and the mitochondrial calcium uniporter (MCU) in IMM represents the main mitochondrial Ca^{2+} uptake pathway. The zoom box shows the electron transport by the respiratory complexes I-IV associated with H^+ transfer from the matrix to the intermembrane space giving rise an electrochemical proton gradient, which in turn is used to the release of ATP synthesized in the catalytic sites of ATP synthase. Created with BioRender.com

The electron flux through the redox centers of ETC is accompanied of H^+ transfer from the mitochondrial matrix to the intermembrane space in complex I, III, and IV, which results in an electrochemical gradient, since IMM is impermeable to H^+ [79]. The electrochemical gradient produces a mitochondrial membrane potential (DY) about −180 to −220 mV, depending on the mitochondrial metabolic (respiratory) state [15]. Then, protons accumulated in the intermembrane space as a result of ETC electron transfer return to the mitochondrial matrix through the F_oF_1-ATP synthase, also known as complex V. The influx of H_+ promotes the rotation of portion F_o that induces conformational alterations of the portion F_1 of ATP synthase, which releases the ATP formed in its catalytic sites, according to the chemiosmotic theory proposed by Peter Mitchell and the conformational hypothesis of ATP synthesis proposed by Paul Boyer [9, 55], as illustrated in the Fig. 6.1. The electron transport with the establishment of an electron chemical gradient (DY) and the ATP synthesis are tightly coupled. Depending on the cells and the physiological and metabolic requirements, such H^+ influx might occur by the uncoupling protein (UCP), promoting the dissipation of the DY to produce heat and/or regulate the production of free radicals by the mitochondria as further discussed [63].

Substrate availability, metabolic regulation, ionic unbalance, and other situations are able to impair the mitochondrial energetic functions. Also, several compounds (chemicals, drugs, xenobiotics, etc.) are able to interfere with the OXPHOS, either through uncoupling or inhibition of the respiratory chain [22, 82]. These effects on mitochondrial energetics can impair physiological functions, being involved in some diseases, either with a harmful or beneficial outcome, as observed in Alzheimer's disease [71] or cancer [65]. Uncoupling of OXPHOS is defined as a dissociation between DY (electron chemical gradient) and ATP synthesis by F_oF_1-ATP synthase, stimulating the state 4 basal respiration, accelerating electron transport and consequently O_2 consumption without ADP phosphorylation to produce ATP, as abovementioned by UCP. Uncoupling influence not only the H^+ leak, but all other mitochondrial functions dependent on DY, energy production, ionic homeostasis, free radical formation, and cell death. Carbonyl cyanide 3-chloro phenylhydrazone (CCCP), carbonyl cyanide 4-(trifluoromethoxy) phenylhydrazone (FCCP), 2,4-dinitrophenol (DNP), and valinomycin (a K^+ ionophore) are well-known examples of mitochondrial uncouplers. Conversely, inhibitors of the respiratory chain interfere with the respiration coupled to the ATP synthesis by acting specifically on respiratory complexes I-V. For example, rotenone blocks the transfer of electrons from NADH to ubiquinone (complex I), malonate inhibits succinate dehydrogenase (complex II), antimycin A blocks the transfer of electrons from ubiquinone to cytochrome *c* (complex III), and cyanide blocks the reduction of O_2 by cytochrome aa3 (complex IV). Importantly, it was shown that nitric oxide (˙NO), an important signaling molecule that plays several physiological functions, inhibits the complex IV (cytochrome *c* oxidase) by competition with the O_2 binding site in a reversible [2], leading to the dissipation of (DY) and therefore diminishing the synthesis of ATP [10]. Other compounds such as fluoroacetate, which inhibits TCA with a consequent reduction in the generation of reduced cofactors, indirectly inhibit OXPHOS. Also, atractyloside, a plant-derived toxic glycoside, inhibits the ADP/ATP exchange by adenine nucleotide translocase (ANT). Finally, F_oF_1-ATP synthase inhibitors, such as oligomycin, inhibits OXPHOS by binding to the F_o portion of the enzyme. The inhibition of the respiratory chain is associated with increased reactive oxygen species (ROS) formation due to the maintenance of the redox centers of complex I-IV fully reduced in the steady-state, resulting in electron leak and incomplete reduction of O_2 to H_2O [61].

Calcium (Ca^{2+}) is involved in the metabolism and energy production, since it acts as a cofactor in several enzymatically-catalyzed metabolic reactions [52]. Also, it is well described the stimulation of OXPHOS by Ca^{2+} [37]. Indeed, Ca^{2+} is a ubiquitous intracellular second messenger and it is involved in several cellular processes and maintenance of cellular Ca^{2+} homeostasis is crucial to the proper functioning [5]. The gradient of Ca^{2+} across the plasma membrane (extra *versus* intracellular compartments) is about 10,000 times, which is maintained by a complex system of channels, transporters, and pumps [14]. Cytosolic Ca^{2+} concentrations are also controlled by the release and uptake by organelles, with emphasis to the endoplasmic reticulum (ER) and also mitochondria. The efflux of Ca^{2+} from the mitochondrial matrix to the cytosol is attributed to the activity of $Na^+/Ca^{2+}/Li^+$ (NCLX) [59] and Ca^{2+}/H^+ antiporters [58]. Conversely, mitochondria have the ability of buffering increases in cytosolic Ca^{2+} levels, which is a relevant mechanism to the maintenance of Ca^{2+} homeostasis and proper cell functioning. Mitochondrial Ca^{2+} influx occurs electrophoretically, i.e., dependent on the DY, by the mitochondrial calcium uniporter (MCU) present in the IMM [4, 36]. However, when this buffering exceeded the organelle capacity, mitochondria undergo a rapid membrane permeabilization, releasing Ca^{2+} and also matrix content to the cytosol. Thus, under several stimuli, mitochondria can undergo a transition in their permeability due to the opening of the mitochondrial permeability transition pore (mPTP), which is formed by different proteins of IMM and OMM, in a process known as mitochondrial permeability transition (MPT). During MPT, it is also observed Ca^{2+} efflux from the matrix to the cytosol, which is associated to cell death [81]. Although the molecular nature of mPTP remains to be completely elucidated, MPT seems to be a Ca^{2+}-dependent process, which is associated with the oxidation of thiol groups of mitochondrial membrane proteins by free radicals or directly by drugs [19, 26, 56, 67]. Additionally, mitochondrial permeabilization might be elicited by the activation of BAX/BAK, proapoptotic members of Bcl-2 family proteins, which oligomerizes in the OMM promoting the outer membrane permeabilization (MOMP) and apoptotic cell death [70]. Although described as a transitory process that can be observed under physiological conditions, mitochondrial permeabilization, which is often associated with cell death [6].

Mitochondrial ROS and Antioxidant Defense System

The execution of mitochondrial respiration is associated with the production of free radicals by mitochondria. These products formed in ETC have been considered harmful due to their capacity to oxidize lipids, proteins, and DNA, leading to a condition known as oxidative stress. Besides the oxidative damage promoted by increased free radical production and/or the failure of the antioxidant defense system, these molecules also might contribute to the maintenance of physiological homeostasis by modulating of signaling pathways and transcription factors [11, 29, 48]. In both situations, free radicals and the alterations in this redox signaling has been associated with aging and diseases.

In order to shed light on this issue, key aspects of mitochondrial free radical formation and elimination will be briefly examined here. It has been proposed that O_2 produced by cyanobacteria started to accumulate in the Earth's atmosphere about 2 billion years ago, which could be toxic to anaerobic eukaryotic cells. According to the Margulis' theory of the endosymbiotic origin of organelles, respiring bacteria invaded unicellular eukaryotic cells enabling them to use O_2 in their energy metabolism. These bacteria would have become the primordial mitochondria [50]. However, despite the energetic advantage in terms of ATP yielding by the oxidation of glucose in aerobic conditions, the utilization of O_2 resulted in the generation of toxic by-products, known as reactive oxygen species (ROS). In fact, mitochondria are the major site of intracellular ROS production uninterruptedly. Under physiological conditions, a small fraction of O_2 undergoes an incomplete reduction during the electron transport by the respiratory chain, being converted into the radical anion superoxide ($O_2^{\cdot -}$) [8, 24]. Such production occurs in the complexes I and III, and also in the ubiquinone cycle. The $O_2^{\cdot -}$ is promptly transformed into hydrogen peroxide (H_2O_2) by the enzymatic action of superoxide dismutases (SOD), being a manganese-dependent superoxide dismutase (Mn-SOD or SOD2) in the mitochondrial matrix or a cooper/zinc-dependent SOD (Cu/Zn-SOD or SOD1) in the cytosol [9, 31].

These species are promptly eliminated by an efficient antioxidant defense system, which is formed by enzymes and non-enzymatic components working cooperatively [44], as depicted in the Fig. 6.2. As abovementioned, $O_2^{\cdot -}$ is converted in H_2O_2 by the peroxidase activity of catalase (CAT) in the cytosol or by the glutathione peroxidase (GPX) in the mitochondria. The reducing equivalents for this reduction are provided by reduced glutathione (GSH) that is converted in its oxidized form GSSG. The rescue of GSH is coupled to the oxidation of NADPH to $NADP^+$, in a reaction catalyzed by glutathione reductase (GR). The reoxidation of $NADP^+$ to provide NADPH to this antioxidant system occurs linked to the glucose metabolism through the pentose phosphate pathway (PPP) with the participation of gluocose-6-phosphate dehydrogenase (G6PD). Although H_2O_2 is not a free radical per se, it is a strong oxidizing agent and under some conditions it might be converted into hydroxyl radical ($^{\cdot}OH$) in a reaction catalyzed by Fe^{2+} or Cu^+, known as Fenton's reaction. Different from $O_2^{\cdot -}$ and H_2O_2, $^{\cdot}OH$ cannot be eliminated by a specific enzymatic antioxidant system and due to its very high reactivity, it usually oxidizes biological targets inside or very close its formation site.

Despite the efficiency of antioxidant defense system to guarantee the cellular homeostasis and perfect functioning of living organisms, oxidative damage can be detected under different conditions, which are associated with excessive ROS production and/or inefficient removal of ROS by the antioxidant system. This condition is known as oxidative stress and it is associated with oxidative damage in biological macromolecules (proteins, lipids, and DNA), impairment of normal organellar and cellular functions, and cell death. The correlation between ROS production and accumulation of age-associated oxidative damage to macromolecules has been proposed over time. Energy metabolism in the mitochondria is linked to the production of ROS.

Mitochondrial (dys)Functions and Aging

Aging is defined as a complex, multifactorial, and progressive process that results in slow functional decline of the physiological functions, which increases the vulnerability to diseases and the probability of death [46]. This broad term, used to describe

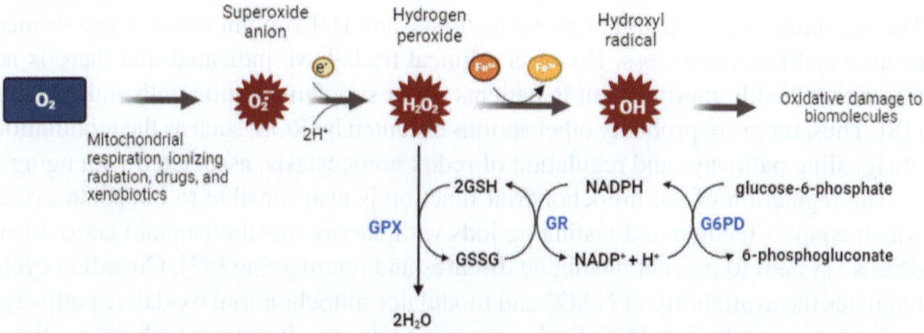

Fig. 6.2 Mitochondrial ROS production and the antioxidant defense system. The superoxide anion ($O_2^{\cdot -}$) is formed by a one-electron reduction of O_2, which is converted in hydrogen peroxide (H_2O_2) by superoxide dismutases (SOD), which is reduced to H_2O by the action of glutathione peroxidase (GPX). Reduced glutathione (GSH) oxidized during H_2O_2 reduction is recovered by the reducing power of NADPH in a reaction catalyzed by glutathione reductase (GR). NADPH is restored by the action of glucose-6-phosphate dehydrogenase (G6PD, pentose phosphate pathway). If not properly eliminated, H_2O_2 can suffer a homo or heterolytic fission catalyzed by transition metals (ferrous Fe^{2+} or cuprous Cu^+ ions) generation the extremely reactive hydroxyl radicals (·OH), prone to oxidize virtually any close biomolecule (proteins, lipids, and nucleic acids). Created with BioRender.com

a plethora of time-related alterations associated with the collapse of physiological functions, seems to be associated with accumulation of damage over time. Several theories of aging have been proposed to try to explain the ongoing aging process, and one of them consider that, at the biological level, ageing results from the accumulative impact of a wide variety of molecular and cellular damages over time [42]. Although it is not clear whether aging acts as causative agent of age-related chronic diseases, certainly it is one of the most important risk factors to be considered in several age-related diseases, including cancer, metabolic syndrome, diabetes mellitus, and neurodegenerative diseases [57]. Anyway, the increasing incidence of chronic diseases has impacted health and the healthcare system [32]. Several molecular changes are detectable during aging. However, it is not also possible to distinguish whether they are cause or consequence of aging [21]. Despite this, a correlation has been proposed between aging and mitochondria, involving the decline of energy mitochondrial metabolism and ROS production.

The free radical theory of aging suggests that the accumulation of oxidative damage of biological macromolecules, compromising cellular and tissue function, is the cause of aging. This was postulated by Harman over sixty years ago [38], evidencing ROS production as a central point of the theory [39]. Another proposal stated that aging could be specifically caused by the cumulative oxidative damage of mitochondrial DNA (mtDNA), resulting in the mitochondrial theory of aging [54]. However, there are several studies corroborating and refuting both theories. Furthermore, studies in *C. elegans* have suggested that ROS, at low levels, might be promotes longevity, as discussed later. It is a consensus that oxidative stress is a component to be considered in aging and age-related diseases, including neurodegenerative diseases.

The accumulative oxidation of proteins, lipids, and DNA might result in age-related decline and functional loss. However, clinical trials have indicated that there is no significant health improvement in patients under supplementation with antioxidants [18]. Thus, there are probably other actions executed by ROS, such as the modulation of signaling pathways and regulation of redox homeostasis, associated with aging.

The regulation of the mitochondrial function is also sensible to circadian cycle, which couples feeding and fasting periods with energy metabolism and antioxidant defense system to prevent metabolic diseases and retard aging [53]. Circadian cycle regulates the availability of NAD^+ and modulates mitochondrial oxidative pathways and cellular metabolism [60]. Such circadian variation altering mitochondrial functions and energy metabolism is associated with aging and extended lifespan [34]. There is an intrinsic difficult to study aging in humans due to average longevity of this specie. Then, other models have been employed to address the molecular mechanisms involved in aging and lifespan, including rats/mice, flies (*Drosophila melanogaster*), zebrafish (*Danio rerio*), and worms (*Caenorhabditis elegans*) [12]. These short-lived species, as *C. elegans*, have an additional advantage by presenting few and well-known genes, with possibility to be genetically easily manipulated. Therefore, these models have been extensively used in aging and longevity studies.

To investigate the role of mitochondrial energetics and ROS production in aging, mutational and knockdown studies were conducted in *C. elegans*. The knockdown of ETC proteins and ATP synthase resulted in increased lifespan, effect that was not achieved by the inhibition of respiratory chain in adulthood [23]. The mutation in the iron sulfur protein (isp-1) of mitochondrial complex III of *C. elegans*, resulted in low O_2 consumption, decreased sensitivity to ROS, and increased life span [27]. Also, the partial impairment of genes encoding complex I (nuo-6) and complex III (isp-1) of the respiratory chain of *C. elegans* also increased longevity apparently by a different mechanism when interference RNA was used to silent these proteins. Interestingly, nuo-6 and isp-1 mitochondrial mutants exhibited increased production of $O_2^{\cdot -}$, without concomitant increase in overall ROS levels and oxidative stress. Yet, the addition of N-acetyl cysteine (NAC) and vitamin C as antioxidants abolished the effect on longevity [76, 77]. And a question that remains to be elucidated is what are the redox signaling alterations induced by $O_2^{\cdot -}$ in *C. elegans* that account for the observed extended longevity. Moreover, clk-1 mutant worms, a gene encoding a hydroxylase involved in the synthesis of ubiquinone, have their lifespan increased by deletion of sod2 gene, but not sod1 or sod5 [68]. Additionally, the effects of $O_2^{\cdot -}$ on lifespan was dependent on the compartment where it was generated, i.e., increased mitochondrial production of ROS extended lifespan, whereas increased cytoplasmic ROS had the opposite effect.

Recently, the concept of hormesis, widely used in Toxicology, has been applied to aging studies [51]. According to it, the response of cells to a high dose harmful/ toxic stimulus might be softened by the previous exposition to low doses of it. In aging, low levels of ROS could enhance cellular fitness and lifespan due to activation homeostatic responses [3]. In fact, such observations have already been done involving the exposition of cardiomyocytes/animals to ischemia/reperfusion (I/R). The I/R damage in heart is associated with oxidative stress and mitochondria occupy a central role in this process by generating high ROS levels. Also, the elimination of damaged mitochondria seems to be important for a positive outcome [80]. In this regard, ischemic preconditioning, i.e., the short and controlled exposure of the cells to intermittent hypoxia, results in an attenuated oxidative damage and a better outcome [25, 72]. Therefore, besides ischemic preconditioning, cardioprotective interventions based on the modulation mitochondrial functions by drugs, specific mitochondrial antioxidants and uncouplers, have been considered as pharmacological approaches to protect mitochondria and cardiovascular function, associated to enhanced longevity [45].

Caloric restriction (CR) represents a reduction in the caloric intake without altering the intake of essential nutrients, such as vitamins and minerals. It has been shown that CR decreases body temperature, indicating the rate of energy metabolism, glucose and lipid blood levels, and ROS production [40, 62], although a prolonged CR to maintain body weight might lead to an undesirable decrease in energy expenditure [47]. Furthermore, CR has been associated with extended lifespan not only in mammals, but also in yeast, worms, flies, and other organisms [13, 28, 75]. The molecular mechanisms leading to these energy metabolism and redox homeostasis alterations are related to modification of mitochondrial bioenergetics, marked by an increased respiratory rate, i.e., a higher electron transport rate and oxygen consumption by the mitochondrial respiratory complexes, which maintain the steady state of respiratory redox centers preferentially in the oxidized form, decreasing electron leak and ROS production. This is also associated with increased consumption of glucose, fatty acids, and other metabolic substrates, with leads to a weight (and fat) loss with improved metabolic condition. Since these alterations might contribute to ameliorate aging-associated deterioration and age-related diseases, several studies have been conducted to find possibilities and strategies with therapeutic potential.

In this regard, intermittent fasting, a particular type of CR, in which periods of regular feeding is alternated with no caloric intake, has been employed as a nutritional approach to treat obesity and metabolic dysfunctions [73]. It has been proposed the use of CR mimetics, i.e., pharmacologically active compounds able to promote the beneficial alterations achieved by CR, which includes the improvement of health and lifespan and the ameliorating aging-related diseases [41]. Different substances are under investigation with different mechanisms of action, i.e., addressing different targets, including intestinal absorption, insulin signaling, epigenetics, inhibitors of glucose absorption, and inhibitors of central executioners of signaling pathways (AMPK, mTOR, sirtuins, and others) [35, 49]. Thus, although CR mimetics have been considered and antiaging elixirs, a better comprehension of all molecular aspects involved and the long-term effects of the use of these substances are required to attest safety and efficacy against aging and age-related cancer and cardiovascular and neurodegenerative diseases. In some aspects, uncoupling of OXPHOS also mimics CR. The uncoupling induced by increased expression of UCP2 in neurons of adult flies increased the respiration rate and decreased ROS production, which was associated to extended lifespan [30]. Interestingly, the chemically-induced uncoupling of OXPHOS was used as strategy to mimic caloric restriction in mice. The protonophore uncoupler 2,4-dinitrophenol increased the respiratory rates and decrease of ROS levels, with was associated with decreased oxidative damage of macromolecules and increased longevity [69]. In isolated brain mitochondria, CR increased their ability to retain Ca^{2+}, named as calcium retention capacity (CRC), without undergoing permeabilization and release Ca^{2+} buffered [1]. Interestingly, the MPT inhibitor cyclosporin A (CsA) increased CRC in brain mitochondria from rats fed ad libitum, similar to observed in CR group, showing that CR decreased the susceptibility of mitochondria to MPT. In fact, the opening of PTP has been proposed as one of the integrating signals between ROS production and aging [66].

Mitochondrial Ca^{2+} handling is strictly controlled by influx and efflux mechanisms. Ca^{2+} regulates mitochondrial metabolism, OXPHOX, and redox homeostasis in mitochondria, but OXPHOS also regulates mitochondrial Ca^{2+} homeostasis, since its uptake depends on DY. Unbalance on these pathways results in mitochondrial matrix Ca^{2+} overload, increased ROS production, and triggering of the mitochondrial permeabilization, which is accompanied of matrix swelling, dissipation of DY, collapses of ATP synthesis, release of proapoptotic proteins, and cell death [74]. All these mitochondrial alterations are tied to each other and they have been obviously associated with several diseases, but also with aging and age-related diseases.

Conclusion

The free radical theory of aging is not sufficient to explain the involvement of mitochondria in aging, because mitochondria execute many processes and ROS production in only one that is tied to the others. Also, ROS might activate protective pathways slowing aging and extending lifespan The notion that oxidative stress, induced by exaggerated ROS formation or decline of the antioxidant system function, causes nonselective and unregulated damage over time and this results in aging is so simple for explain a multifaceted and complex process. Thus, there are strong evidence point to the involvement of mitochondria in aging, but all mitochondrial processes and functions are probably involved, not only ROS production. Therefore, better characterization of the molecular mechanisms involved in the triggering and regulation of the redox signaling and homeostasis, as well as the identification of cellular responses to the different stimuli related to aging, to improve the comprehension about how the mitochondrial function and dysfunctions contributes to aging process and also to the aging-related diseases.

Acknowledgements I would like to thank Ms. Vyctoria dos Santos Ramos (M.Sc.) from Federal University of ABC for your valuable technical assistance with the figures that illustrated this chapter and also to the Fundação de Amparo à Pesquisa do Estado de São Paulo (FAPESP) for the financial support (2021/14650-3).

References

1. Amigo I, Menezes-Filho SL, Luévano-Martínez LA, Chausse B, Kowaltowski AJ (2017) Caloric restriction increases brain mitochondrial calcium retention capacity and protects against excitotoxicity. Aging Cell 16(1):73–81. doi: https://doi.org/10.1111/acel.12527
2. Antunes F, Boveris A, Cadena E (2004) On the mechanism and biology of cytochrome oxidase inhibition by nitric oxide. Proc Natl Acad Sci USA 101:16774–16779. doi: https://doi.org/10.1073/pnas.0405368101
3. Barbosa MC, Grosso RA, Fader CM (2018) Hallmarks of aging: an autophagic perspective. Front Endocrinol 9:790. doi: https://doi.org/10.1002/pmic.201800400
4. Baughman JM, Perocchi F, Girgis HS, Plovanich M, Belcher-Timme CA, Sancak Y, Bao XR, Strittmatter L, Goldberger O, Bogorad RL, Koteliansky V, Mootha VK (2011) Integrative genomics identifies MCU as an essential component of the mitochondrial calcium uniporter. Nature 476:341–345. doi: https://doi.org/10.1038/nature10234
5. Berridge MJ, Lipp P, Bootman MD (2000) The versatility and universality of calcium signalling, Nat Rev Mol Cell Biol 1:11–21. doi: https://doi.org/10.1038/35036035
6. Bock FJ, Tait SWG (2020) Mitochondria as multifaceted regulators of cell death. Nat Rev Mol Cell Biol 21(2):85–100. doi: https://doi.org/10.1038/s41580-019-0173-8
7. Boveris A, Cadenas E, Stoppani AOM (1976) Role of ubiquinone in the mitochondrial generation of hydrogen peroxide. Biochem J 156:435. doi: https://doi.org/10.1042/bj1560435
8. Boveris A, Oshino N, Chance B (1972) The cellular production of hydrogen peroxide. Biochem J 128:617–630. doi: https://doi.org/10.1042/bj1280617
9. Boyer PD, Cross RL, Momsen W (1973) A new concept for energy coupling in oxidative phosphorylation based on a molecular explanation of the oxygen exchange reactions. Proc Natl Acad Sci USA 70:2837–2839. doi: https://doi.org/10.1073/pnas.70.10.2837
10. Brookes PS, Bolanos JP, Heales SJ (1999) The assumption that nitric oxide inhibits mitochondrial ATP production is correct. FEBS Lett 446:261–263. doi: https://doi.org/10.1016/s0014-5793(99)00217-3
11. Brookes PS, Levonen AL, Shiva S, Sarti P, and Darley-Usmar VM (2002) Mitochondria: regulators of signal transduction by reactive oxygen and nitrogen species. Free Radic Biol Med 33:755–764. doi: https://doi.org/10.1016/s0891-5849(02)00901-2
12. Brunet A (2020) Old and new models for the study of human ageing. Nat Rev Mol Cell Biol 21(9):491–493. doi: https://doi.org/10.1038/s41580-020-0266-4.
13. Burger JMS, Buechel SD, Kawecki TJ (2010) Dietary restriction affects lifespan but not cognitive aging in *Drosophila melanogaster*. Aging Cell 9:327–335. doi: https://doi.org/10.1111/j.1474-9726.2010.00560.x
14. Carafoli E (1988) Membrane transport of calcium: an overview. Methods Enzymol 157:3–11. doi: https://doi.org/10.1016/0076-6879(88)57063-5.
15. Chance B, Williams GR (1956) The respiratory chain and oxidative phosphorylation. Adv Enzymol Relat Subj Biochem 17:65–134. doi: https://doi.org/10.1002/9780470122624.ch2
16. Chandel NS (2015) Evolution of mitochondria as signaling organelles. Cell Metab 22:204–206 (2015). doi: https://doi.org/10.1016/j.cmet.2015.05.013
17. Christensen SB (2021) Drugs That Changed Society: History and Current Status of the Early Antibiotics: Salvarsan, Sulfonamides, and β-Lactams. Molecules. 26(19):6057. doi: https://doi.org/10.3390/molecules26196057
18. Conti V, Izzo V, Corbi G, Russomanno G, Manzo V, De Lise F, Di Donato A, Filippelli A (2016) Antioxidant Supplementation in the Treatment of Aging-Associated Diseases. Front Pharmacol 7:24. doi: https://doi.org/10.3389/fphar.2016.00024
19. Cruz TS, Faria PA, Santana DP, Ferreira JC, Oliveira V, Nascimento OR, Cerchiaro G, Curti C, Nantes IL, Rodrigues T (2010) On the mechanisms of phenothiazine-induced mitochondrial permeability transition: Thiol oxidation, strict Ca^{2+} dependence, and cyt *c* release. Biochem Pharmacol 80(8):1284–1295. doi: https://doi.org/10.1016/j.bcp.2010.06.052

20. Cunningham CN, Rutter J (2020) 20,000 picometers under the OMM: diving into the vastness of mitochondrial metabolite transport. EMBO Rep 21(5):e50071. doi: https://doi.org/10.15252/embr.202050071
21. Dancy BM, Sedensky MM, Morgan PG (2014) Effects of the mitochondrial respiratory chain on longevity in *C. elegans*. Exp Gerontol 56:245–255. doi: https://doi.org/10.1016/j.exger.2014.03.028
22. Demine S, Renard P, Arnould T (2019) Mitochondrial uncoupling: a key controller of biological processes in physiology and diseases. Cells 8(8):795. doi: https://doi.org/10.3390/cells8080795
23. Dillin A, Hsu A-L, Arantes-Oliveira N, Lehrer-Graiwer J, Hsin H, Fraser AG, Kamath RS, Ahringer J, Kenyon C (2002) Rates of behavior and aging specified by mitochondrial function during development. Science 298(5602):2398–2401. doi: https://doi.org/10.1126/science.1077780
24. Dröse S, Brandt U (2012) Molecular mechanisms of superoxide production by the mitochondrial respiratory chain. Adv Exp Med Biol 748:145–169. doi: https://doi.org/10.1007/978-1-4614-3573-0_6
25. Facundo HTF, Carreira RS, de Paula JG, Santos CCX, Ferranti R, Laurindo FRM, Kowaltowski AJ (2006) Ischemic preconditioning requires increases in reactive oxygen release independent of mitochondrial K^+ channel activity. Free Radic Biol Med 40:469–479. doi: https://doi.org/10.1016/j.freeradbiomed.2005.08.041
26. Fagian MM, Pereira-da-Silva L, Martins IS, Vercesi AE (1990) Membrane protein thiol cross-linking associated with the permeabilization of the inner mitochondrial membrane by Ca^{2+} plus prooxidants. J Biol Chem 265(32):19955–19960.
27. Feng J, Bussière F, Hekimi S (2001) Mitochondrial electron transport is a key determinant of life span in *Caenorhabditis elegans*. Dev Cell 1(5):633–644. doi: https://doi.org/10.1016/s1534-5807(01)00071-5
28. Fontana L, Partridge L, Longo VD (2010) Extending healthy life span--from yeast to humans. Science 328(5976):321–326. doi: https://doi.org/10.1126/science.1172539
29. Forman HJ (2016) Redox signaling: An evolution from free radicals to aging. Free Radic Biol Med 97:398–407. doi: https://doi.org/10.1016/j.freeradbiomed.2016.07.003
30. Fridell YW, Sánchez-Blanco A, Silvia BA, Helfand SL (2005) Targeted expression of the human uncoupling protein 2 (hUCP2) to adult neurons extends life span in the fly. Cell Metab 1:145–152. doi: https://doi.org/10.1016/j.cmet.2005.01.005
31. Fridovich I (1975) Superoxide dismutases. Annu Rev Biochem 44:147–159. doi: https://doi.org/10.1146/annurev.bi.44.070175.001051
32. Gaugler JE (2015) Chronic disease and aging: screening and management applications. J Appl Gerontol 34(3):273–276. doi: https://doi.org/10.1177/0733464815573392
33. Giacomello M, Pyakurel A, Glytsou C, Scorrano L (2020) The cell biology of mitochondrial membrane dynamics. Nat Rev Mol Cell Biol 21(4):204–224. doi: https://doi.org/10.1038/s41580-020-0210-7
34. de Goede P, Wefers J, Brombacher EC, Schrauwen P, Kalsbeek A (2018) Circadian rhythms in mitochondrial respiration. J Mol Endocrinol 60(3):R115–R130. doi: https://doi.org/10.1530/JME-17-0196.
35. Green CL, Lamming DW, Fontana L (2022) Molecular mechanisms of dietary restriction promoting health and longevity. Nat Rev Mol Cell Biol 23(1):56–73. doi: https://doi.org/10.1038/s41580-021-00411-4
36. Gunter TE, Pfeiffer DR (1990) Mechanisms by which mitochondria transport calcium. Am J Physiol 258:C755–786. doi: https://doi.org/10.1152/ajpcell.1990.258.5.C755
37. Hansford RG, Zorov D (1998) Role of mitochondrial calcium transport in the control of substrate oxidation. Mol Cell Biochem 184:359–369
38. Harman D (1956) Aging: a theory based on free radical and radiation chemistry. J Gerontol 11:298–300. doi: https://doi.org/10.1093/geronj/11.3.298
39. Harman D (1972) The biologic clock: the mitochondria? J Am Geriatr Soc 20:145–147. doi: https://doi.org/10.1111/j.1532-5415.1972.tb00787.x

40. Heilbronn LK, Jonge L, Frisard MI, DeLany JP, Larson-Meyer DE, Rood J, Nguyen T, Martin CK, Volaufova J, Most MM, Greenway FL, Smith SR, Deutsch WA, Williamson DA, Ravussin E, Pennington CALERIE Team (2006) Effect of 6-month calorie restriction on biomarkers of longevity, metabolic adaptation, and oxidative stress in overweight individuals: a randomized controlled trial. JAMA 295:1539–1548. doi: https://doi.org/10.1001/jama.295.13.1539
41. Hofer SJ, Davinelli D, Bergmann M, Scapagnini G, Madeo F (2021) Caloric restriction mimetics in nutrition and clinical trials. Front Nutr 8:717343. doi: https://doi.org/10.3389/fnut.2021.717343
42. Jin K (2010) Modern Biological Theories of Aging. Aging Dis 1(2):72–74.
43. Kowaltowski AJ (2000) Alternative mitochondrial functions in cell physiopathology: beyond ATP production. Braz J Med Biol Res 33(2):241–250. doi: https://doi.org/10.1590/s0100-879x2000000200014
44. Kowaltowski AJ, de Souza-Pinto NC, Castilho RF, Vercesi AE (2009) Mitochondria and reactive oxygen species. Free Radic Biol Med 47:333–343. doi: https://doi.org/10.1016/j.freeradbiomed.2009.05.004
45. Kuznetsov AV, Javadov S, Margreiter R, Grimm M, Hagenbuchner J, Ausserlechner MJ (2019) The Role of Mitochondria in the Mechanisms of Cardiac Ischemia-Reperfusion Injury. Antioxidants (Basel) 8(10):454. doi: https://doi.org/10.3390/antiox8100454
46. Labat-Robert J, Robert L (2015) Longevity and aging. Mechanisms and perspectives. Pathol Biol 63(6):272–276. doi: https://doi.org/10.1016/j.patbio.2015.08.001
47. Leibel RL, Rosenbaum M, Hirsch J (1995) Changes in energy expenditure resulting from altered body weight. N Engl J Med 332:621–628. doi: https://doi.org/10.1056/NEJM199503093321001
48. Lennicke C, Cochemé HM (2021) Redox metabolism: ROS as specific molecular regulators of cell signaling and function. Mol Cell 81(18):3691–3707. doi:https://doi.org/10.1016/j.molcel.2021.08.018
49. Madeo F, Carmona-Gutierrez D, Hofer SJ, Kroemer G (2019) Caloric Restriction Mimetics against Age-Associated Disease: Targets, Mechanisms, and Therapeutic Potential. Cell Metab 29(3):592–610. doi: https://doi.org/10.1016/j.cmet.2019.01.018
50. Margulis L, Bermudes D (1985) Symbiosis as a mechanism of evolution: status of cell symbiosis theory. Symbiosis 1:101–124.
51. Mattson MP (2008) Hormesis defined. Ageing Res Rev 7(1):1–7. doi: https://doi.org/10.1016/j.arr.2007.08.007
52. McCormack JG, Denton RM (1993) Mitochondrial Ca^{2+} transport and the role of intramitochondrial Ca^{2+} in the regulation of energy metabolism. Dev Neurosci 15:165–173. doi: https://doi.org/10.1159/000111332
53. Mezhnina V, Ebeigbe OP, Poe A, Kondratov RV (2022) Circadian Control of Mitochondria in Reactive Oxygen Species Homeostasis. Antioxid Redox Signal 37(10–12):647–663. doi: https://doi.org/10.1089/ars.2021.0274.
54. Miquel J, Economos AC, Fleming J, Johnson Jr JE (1980) Mitochondrial role in cell aging. Exp Gerontol 15:575–591. doi: https://doi.org/10.1016/0531-5565(80)90010-8
55. Mitchell P (1961) Coupling of phosphorylation to electron and hydrogen transfer by a chemi-osmotic type of mechanism. Nature 191:144–148. doi: https://doi.org/10.1038/191144a0
56. Moraes VW, Caires AC, Paredes-Gamero EJ, Rodrigues T (2013) Organopalladium compound 7b targets mitochondrial thiols and induces caspase-dependent apoptosis in human myeloid leukemia cells. Cell Death Dis 4(6):e658. doi: https://doi.org/10.1038/cddis.2013.190
57. Niccoli T, Partridge L (2012) Ageing as a risk factor for disease. Curr Biol 22:R741–R752. doi: https://doi.org/10.1016/j.cub.2012.07.024
58. Nicholls DG, Crompton M (1980) Mitochondrial calcium transport. FEBS Lett 111:261–268. doi: https://doi.org/10.1016/0014-5793(80)80806-4
59. Palty R, Silverman WF, Hershfinkel M, Caporale T, Sensi SL, Parnis J, Nolte C, Fishman D, Shoshan-Barmatz V, Herrmann S, Khananshvili D, Sekler I (2010) Nclx is an essential component of mitochondrial Na^{+}/Ca^{2+} exchange. Proc Natl Acad Sci USA 107:436–441. doi: https://doi.org/10.1073/pnas.0908099107

60. Peek CB, Affinati AH, Ramsey KM, Kuo HY, Yu W, Sena LA, Ilkayeva O, Marcheva B, Kobayashi Y, Omura C, Levine DC, Bacsik DJ, Gius D, Newgard CB, Goetzman E, Chandel NS, Denu JM, Mrksich M, Bass J (2013). Circadian clock NAD^+ cycle drives mitochondrial oxidative metabolism in mice. Science 342(6158):1243417. doi: https://doi.org/10.1126/science.1243417
61. Quinlan CL, Gerencser AA, Treberg JR, Brand MD (2011) The mechanism of superoxide production by the antimycin-inhibited mitochondrial Q-cycle. J Biol Chem 286:31361–72. doi: https://doi.org/10.1074/jbc.M111.267898
62. Redman LM, Smith SR, Burton JH, Martin CK, Il'yasova D, Ravussin E (2018) Metabolic slowing and reduced oxidative damage with sustained caloric restriction support the rate of living and oxidative damage theories of aging. Cell Metab. 2018; 27: 805–815.e4. DOI: https://doi.org/10.1016/j.cmet.2018.02.019
63. Ricquier D, Bouillaud F (2000) Mitochondrial uncoupling proteins: from mitochondria to the regulation of energy balance. J Physiol 529(1):3–10. doi: https://doi.org/10.1111/j.1469-7793.2000.00003.x
64. Rodrigues T, Ferraz LS (2020) Therapeutic potential of targeting mitochondrial dynamics in cancer. Biochem Pharmacol 182:114282. doi: https://doi.org/10.1016/j.bcp.2020.114282
65. Rohlena J, Dong LF, Ralph SJ, Neuzil J (2011) Anticancer drugs targeting the mitochondrial electron transport chain. Antioxid Redox Signal 15(12):2951–74. doi: https://doi.org/10.1089/ars.2011.3990
66. Rottenberg H, Hoek JB (2017) The path from mitochondrial ROS to aging runs through the mitochondrial permeability transition pore. Aging Cell 16(5):943–955. doi: https://doi.org/10.1111/acel.12650
67. Santana DP, Faria PA, Paredes-Gamero EJ, Caires AC, Nantes IL, Rodrigues T (2009) Palladacycles catalyse the oxidation of critical thiols of the mitochondrial membrane proteins and lead to mitochondrial permeabilization and cytochrome c release associated with apoptosis. Biochem J 417(1):247–256. doi: https://doi.org/10.1042/BJ20080972
68. Schaar CE, Dues DJ, Spielbauer KK, Machiela E, Cooper JF, Senchuk M, Hekimi S, Van Raamsdonk JM (2015) Mitochondrial and cytoplasmic ROS have opposing effects on lifespan. PLoS Genet 11(2):e1004972. doi: https://doi.org/10.1371/journal.pgen.1004972
69. Silva CCC, Cerqueira FM, Barbosa LF, Medeiros MHG, Kowaltowski AJ (2008) Mild mitochondrial uncoupling in mice affects energy metabolism, redox balance and longevity. Aging Cell 7(4):552–560. doi: https://doi.org/10.1111/j.1474-9726.2008.00407.x
70. Tait SWG, Green DR (2013) Mitochondrial regulation of cell death. Cold Spring Harb Perspect Biol 5(9): a008706. doi: https://doi.org/10.1101/cshperspect.a008706
71. Trushina E, Trushin S, Hasan MF (2022) Mitochondrial complex I as a therapeutic target for Alzheimer's disease. Acta Pharm Sin B 12(2):483–495. doi: https://doi.org/10.1016/j.apsb.2021.11.003
72. Vanden Hoek T, Becker LB, Shao ZH, Li CQ, Schumacker PT (2000) Preconditioning in cardiomyocytes protects by attenuating oxidant stress at reperfusion. Circ Res 86:541–548. doi: https://doi.org/10.1161/01.res.86.5.541
73. Varady KA, Cienfuegos S, Ezpeleta M, Gabel K (2022) Clinical application of intermittent fasting for weight loss: progress and future directions. Nat Rev Endocrinol 18(5):309–321. doi: https://doi.org/10.1038/s41574-022-00638-x
74. Vercesi AE, Roger F Castilho, Alicia J Kowaltowski, Helena C F de Oliveira, Nadja C de Souza-Pinto, Tiago R Figueira, Estela N B Busanello (2018) Mitochondrial calcium transport and the redox nature of the calcium-induced membrane permeability transition. Free Radic Biol Med 129:1–24. doi: https://doi.org/10.1016/j.freeradbiomed.2018.08.034
75. Wu Z, Isik M, Moroz N, Steinbaugh MJ, Zhang P, Blackwell TK (2019) Dietary restriction extends lifespan through metabolic regulation of innate immunity. Cell Metab 29(5):1192–1205.e8. doi: https://doi.org/10.1016/j.cmet.2019.02.013
76. Yang W, Hekimi S (2010a) A mitochondrial superoxide signal triggers increased longevity in *Caenorhabditis elegans*. PLoS Biol 8(12):e1000556. doi: https://doi.org/10.1371/journal.pbio.1000556

77. Yang W, Hekimi S (2010b) Two modes of mitochondrial dysfunction lead independently to lifespan extension in Caenorhabditis elegans. Aging Cell 9(3):433–447. doi: https://doi.org/10.1111/j.1474-9726.2010.00571.x
78. Zahedi R, Sickmann A, Boehm AM, Winkler C, Zufall N, Schönfisch B, Guiard B, Pfanner N, Meisinger C (2006) Proteomic Analysis of the Yeast Mitochondrial Outer Membrane Reveals Accumulation of a Subclass of Preproteins. Mol Biol Cell 17:1436–1450. doi: https://doi.org/10.1091/mbc.e05-08-0740
79. Zhao R-Z, Jiang S, Zhang L, Yu Z-B (2019) Mitochondrial electron transport chain, ROS generation and uncoupling. Int J Mol Med 44(1):3–15. doi: https://doi.org/10.3892/ijmm.2019.4188
80. Zhou M, Yu Y, Luo X, Wang J, Lan X, Liu P, Feng Y, Jian W (2021) Myocardial ischemia-reperfusion injury: therapeutics from a mitochondria-centric perspective. Cardiology 146(6):781–792. doi: https://doi.org/10.1159/000518879
81. Zoratti M, Szabó I (1995) The mitochondrial permeability transition. Biochim Biophys Acta 1241:139–176. doi: https://doi.org/10.1016/0304-4157(95)00003-a
82. van der Stel W, Yang H, Vrijenhoek NG, Schimming JP, Callegaro G, Carta G, Darici S, Delp J, Forsby A, White A, le Dévédec S, Leist M, Jennings P, Beltman JB, van de Water B, Danen EHJ (2022) Mapping the cellular response to electron transport chain inhibitors reveals selective signaling networks triggered by mitochondrial perturbation. Arch Toxicol 96(1):259–285. doi: https://doi.org/10.1007/s00204-021-03160-7

Chapter 7
Stem Cell Exhaustion

Valquiria Bueno

Ageing and Stem Cells

A great medical burden associated to non-communicable chronic diseases has been predicted because of the global ageing of the population. Individuals over 65 years old likely present one or more chronic diseases that contribute for the multimorbidity observed in this population [1]. Ischemic heart disease, osteoporosis, osteoarthritis, metabolic syndrome, type 2 diabetes, chronic renal disease, chronic lung diseases and Alzheimer's disease are the most prevalent chronic diseases in old adults.

The tissue homeostasis is provided by adult stem cells (rising from embryonic stem cells) which are a reservoir of cells with capacity to generate somatic cell lineages. Unlike embryonic stem cells that proliferate actively, the majority of adult stem cells are in quiescent state and are induced to proliferate for regenerative purposes and repair. Stem cells are susceptible to DNA damage during the quiescent state which leads to stem cells exhaustion due to accumulation of mutations. Moreover, epigenetic modifications (DNA methylation in specific genes) are also associated with stem cell exhaustion [2, 3]. Thus, DNA damage, epigenetic alteration and other changes occurring in stem cells cause decline in cell number and function, being referred as "stem cell exhaustion" which has been considered one major hallmark of ageing [4–7].

Chronic diseases development has been linked to the exhaustion in the repair process and probably as a consequence, a considerable percentage of individuals older than 65 years present multiple morbidities. Important questions are still unanswered such as how much the changes in number and/or functions of stem cells

V. Bueno (✉)
Department of Microbiology Immunology and Parasitology, UNIFESP Federal University of São Paulo, Botucatu St 862 4° Level, São Paulo 04023-900, Brazil
e-mail: vbueno@unifesp.br

V. Bueno (ed.), *Cellular and Molecular Aspects of Ageing*,
https://doi.org/10.1007/978-3-031-55022-5_7

contribute to organ deterioration and what is the role played by stem cells in age-related diseases. On the other hand, some correlations between alterations in stem cells and diseases have been described as it follows:

- the age-associated myelodysplastic syndrome has been linked to hematopoietic stem cells (HSCs) dysfunction (loss of TRAF6 in pre-leukemic cells) that could be exacerbated in an aged inflammatory context [8],
- in dyskeratosis congenita some patients present mutations in an TRF1-interacting protein which when chronically deleted in mice leads to DNA damage and cellular senescence in addition to severe bone marrow aplasia [9],
- the whole-exome sequencing DNA of peripheral-blood cells showed that somatic mutations rise with age and are associated with increased risk for: hematologic cancer, all-cause mortality, incident coronary heart disease, and ischemic stroke [10],
- as HSCs age, they display increased DNA damage and DNA damage response during normal self-renew and differentiation divisions but also in conditions of proliferative or inflammatory stress [11–14].

Considering that stem cell evaluation has technical limitations such as heterogeneity of cells and isolation of a pure stem cell population in adequate numbers, the majority of studies have been performed in neural stem cell, bone marrow stromal stem cells, epithelial stem cells, and hematopoietic stem cells. Some findings are described below:

- Adult neural stem cells (NSC) are responsible for the continuous neurogenesis not only generating new neurons but also maintaining a reserved pool in quiescence probably to supply the demand of new neurons. Moreover, aged mammals with cognitive impairment display reduction in neural stem cells and neurogenesis [15].
- Bone marrow stromal stem cells (BMSC) are responsible for bone tissue remodeling/regeneration and differentiation into osteoblasts and adipocytes [16]. Using skeletal muscle- and bone-derived mesenchymal stem/stromal cells (MSC) from patients with osteoarthritis and osteoporosis, Čamernik et al. found ex vivo and in vitro declined osteogenesis and lower expression of leptin receptor in comparison with healthy individuals [17]. Moreover, older patients' BMSC showed in culture decreased growth, low capacity for osteogenic differentiation and higher capacity for adipocyte differentiation [18].
- Epithelial adult stem cells (ESC) have been associated with self-renewal and potential unlimited. They reside in specific regions of the tissue in quiescence until some stimuli for normal reposition of cells or for response to insults occur [19]. In skin, epithelial stem cells located in the interfollicular epidermis has been linked to the tissue homeostasis in response to the continuous insults (thermal shocks, chemical agents, pathogens) [20]. The same has been found in the oral mucosa as this tissue is responsible for the oral cavity protection (pathogens and mastication-induced stress). For the cornea epithelium regeneration, human limbal epithelial stem cells (H-LESC) are responsible for tissue homeostasis [21].

In primary human keratinocytes isolated from skin, conjuntiva, oral mucosa and limbus biopsies of healthy individuals, Barbaro et al. found that, in vitro, stem cells divisions are tightly regulated (frequency and timing) and specific for each tissue [22].

Hematopoietic Stem Cells (HSC)

The first stem cells isolated and identified were the hematopoietic stem cells which in human are characterized as Lin^-, $CD34^+$, $CD38^-$, $Thy1.1^+$, $CD45RA^-$ [23–25].

HSC are responsible for the replenishment of blood cells (hematopoiesis) and the study of how the age process affects the number and function of HSC can lead to a further understanding of age-related malignancies, age-related changes in the immunity, and possible targets for therapy.

HSC is maintained in quiescence (non-cycling) keeping the requirement of energy low, preventing oxidative stress, and providing a lifelong stable pool of HSC which is activated (if needed) to maintain the hematopoietic system homeostasis. HSC (quiescence, self-renewal, differentiation) is controlled by transcription, epigenetic and metabolism, in addition to cell-cycle regulators. Extrinsically HSC receives signals from cells and extracellular matrix (bone marrow niche) [26–29]. Other signals such infections, pathologies, physiological stress (ageing) can also play a role [30–32]. Thus, HSC proliferation and differentiation can be affected by acute stress with fast resolution, and chronic stress (infection/inflammation, chemotherapy) which drives HSC to exhaustion [reviewed in 33]. Considering that the age process has been associated with chronic stress, it was found that CD34 + HSCs from aged individuals present increased DNA damage and double-strand breaks in comparison with young donors [34].

The decrease of HSC clonal diversity in individuals older than 70 years was reported by Mitchell et al. through sequencing of single-cell-derived colonies of hematopoietic progenitor cells and HSC (3579 genomes) [35]. In humans, the heterogeneity in hematopoietic stem cells leads to clonal hematopoiesis (few HSC clones contributing in a disproportional way to blood production) and competition among stem cells. It is ubiquitous in old individuals and can be due to survival and proliferation advantage of mutant stem cells. Clonal hematopoiesis increases the risk of malignant development and it has been associated with the age-related increased incidence of blood cancer in humans [36–39]. A specific sort of clonal hematopoiesis—clonal hematopoiesis of indeterminate potential (CHIP) or age-related clonal hematopoiesis (ARCH)—has been associated with hematological malignancy, cardiovascular disease, stroke, diabetes, arthritis, liver disease and solid cancer [40]. Clonal hematopoiesis has been linked to genes encoding epigenetics regulators (i.e. DNA methyltransferase 3A) suffering somatic mutations [38, 39] but also to an inflammatory microenvironment characteristic of the ageing process. Aged HSC niche shows upregulation of TNF (tumor necrosis factor), TGF-β (transforming growth factor- β), IL-6, IL-1 [41–44] which are inflammatory cytokines associated with “inflammageing” (extensively discussed—Chap. 2 “Cellular Senescence” and chapter 11 “Senescence-Associated Secretory Process SASP). Another important change in HSC during the ageing process is a bias towards myeloid cells output and decrease of lymphoid cells [45]. Rundberg et al. found that human cells from cord blood to aged bone marrow presented decrease in the frequency of common lymphocyte progenitors (CLP) and increase of megakaryocyte/erythrocyte progenitors (MEP). In addition, they observed intrinsic alterations in phenotype, function, and transcription of human HSC due to ageing [46].

Hematopoietic Stem Cells and Immunity

During the ageing process, changes occurring in both hematopoietic and immune system result in impaired immunity. Older individuals are thus more susceptible to infections, present a less efficient response to vaccines, and display increased risk for anemia, hematologic malignancy, and autoimmunity.

Ageing has been associated with chronic increased levels of circulating TNF-alpha, IL-1beta, IL-6, and other pro-inflammatory factors which can harm HSC [47–49]. In this scenario the expansion of HSC displaying CHIP mutations and development of myeloid malignancy is favored and has been linked to the chronic presence of pro-inflammatory cytokines [50–53].

Receptors of the immune innate response (TLRs and IFN receptors) are expressed by HSC allowing these cells to respond to inflammatory mediators and leading to changes in their proliferation and lineage fate which will affect the immune defense [54–58].

In a lifetime, several factors (physical, chemical, biological) can lead to lymphopenia and in consequence homeostatic proliferation is induced to restore T lymphocytes in periphery. As a result, the selection of clones of autoreactive T-cells is impaired in addition to the failure in clearing infections with chronicity of the infectious process [59–61]. The decrease in the output of naïve lymphocytes causes lymphopenia and augments the harms of homeostatic proliferation [59, 62].

The impairment in lymphoid precursors leads to the decrease in production the new B-cells by the bone marrow and T-cells by the thymus with consequent lower diversity of B- and T-cells receptors [63–65]. As a result there is an increased risk for autoimmune diseases, less efficacy against infections and tumor development [66, 67].

In myeloid cells, ageing has been associated with increased cells numbers and decreased function. Antigen processing and presentation are decreased, there is a lower secretion of interferon, in addition to a diminished recognition and destruction of senescent cells. Thus, innate immune cells present impaired recognition of infectious agents and mutated cells, lower removal of apoptotic and senescent cells, and secretion of pro-inflammatory factors [68–70].

Induced Pluripotent Stem Cells (iPSC)

Regenerative medicine by reprograming somatic stem cells into induced pluripotent stem cells (iPSC) aims to delay the progression of chronic diseases and to develop new therapies. Using cells of the patient itself the reprogramming can be done by the conventional method (Yamanaka's factors—OCT3/4, SOX2, KLF4, and c-MYC) or specific conditions in culture [71, 72].

In age-related macular disease (AMD) it was possible to modulate the early pathogenesis deriving retinal pigment epithelium (RPE) from AMD patients and control donors [73, 74]. There are clinical trials ongoing using iPSC and targeting AMD. In Alzheimer's disease, it has been possible to further understand the pathological mechanisms of the disease using iPSC technology [75–77]. Aiming the treatment of cardiovascular diseases (CVD), iPSC has been differentiated in to myocytes, vascular smooth muscle cells, endothelial cells, and fibroblasts [78–81]. In the cancer field, it has been possible to epigenetically reprogram tumor cells into normal cells, showing thus that are susceptibility from the cancer cells to epigenetic reversion [82]. Organoids derived from iPSC have been used to study renal cancer, glioblastoma, pancreatic cancer, lung cancer, and colorectal cancer [83–87]. Regarding to immunity in cancer, it was possible to generate tumor-specific iPSC from T-lymphocytes infiltrating the tumor in human colorectal cancer. Those iPSC-T cells were capable to proliferate and kill tumor targets suggesting anti-tumor activity that can be used as cancer therapy [88]. In humans, it was shown that iPSC-NK cells presented high antibody-dependent cellular cytotoxicity and were efficient in hematologic and solid tumors [89]. Another possibility is to use iPSC to express cancer cell-engaging proteins as CAR (chimeric antigen receptors) and improve the effector cells activity against tumor. In patients with B-cell malignancies it is in use a CAR-T cell product derived from iPSC in a clinical trial [90, 91].

Acknowledgements CAPES PrInt UNIFESP 88881.310735/2018-01

References

1. Barnett K, Mercer SW, Norbury M. Epidemiology of multimorbidity and implications for health care, research, and medical education: a cross-sectional study. Lancet 2012; 380: 37–43.
2. Beerman I, Seita J, Inlay MA, Weissman IL, Rossi DJ. Quiescent hematopoietic stem cells accumulate DNA damage during aging that is repaired upon entry into cell cycle. Cell Stem Cell 2014; 15(1): 37- 50. doi:https://doi.org/10.1016/j.stem.2014.04. 016
3. Vas V, Senger K, Dörr K, Niebel A, Geiger H. Aging of the micro-environment influences clonality in hematopoiesis. PLoS One 2012; 7(8): e42080. doi: 10.1371/ journal.pone.0042080
4. López-Otín C, Blasco MA, Partridge L, Serrano M, Kroemer G. The hallmarks of aging. Cell 2013; 153:1194–1217.
5. Goodell MA, Rando TA. Stem cells and healthy aging. Science 2015; 350 (6265): 1199–204. doi: https://doi.org/10.1126/science.aab3388.
6. Navarro Negredo P, Yeo RW, Brunet A. Aging and Rejuvenation of Neural Stem Cells and Their Niches. Cell Stem Cell 2020; 27(2):202–223. doi: 10.1016/j. stem.2020.07.002

7. Oh J, Lee YD, Wagers AJ. Stem cell aging: mechanisms, regulators and therapeutic opportunities. Nat Med 2014; 20(8):870–80. doi: https://doi.org/10.1038/nm.3651.
8. Muto T, Guillamot M, Yeung J, Fang J, Bennett J, Nadorp B, et al. TRAF6 functions as a tumor suppressor in myeloid malignancies by directly targeting MYC oncogenic activity. Cell Stem Cell 2022; 29(2):298–314.e9. doi: https://doi.org/10.1016/j.stem.2021.12.007.
9. Beier F, Foronda M, Martinez P, Blasco MA. Conditional TRF1 knockout in the hematopoietic compartment leads to bone marrow failure and recapitulates clinical features of dyskeratosis congenita. Blood 2012; 120(15):2990–3000. doi: https://doi.org/10.1182/blood-2012-03-418038.
10. Jaiswal S, Fontanillas P, Flannick J, Manning A, Grauman PV, Mar BG, et al. Age-related clonal hematopoiesis associated with adverse outcomes. N Engl J Med 2014; 371(26):2488–98. doi: https://doi.org/10.1056/NEJMoa1408617.
11. Yamamoto R, Morita Y, Ooehara J, Hamanaka S, Onodera M, Rudolph KL, et al. Clonal analysis unveils self-renewing lineage-restricted progenitors generated directly from hematopoietic stem cells. Cell 2013; 154(5):1112–1126. doi: 10.1016/j. cell.2013.08.007.
12. Asada N, Takeishi S, Frenette PS. Complexity of bone marrow hematopoietic stem cell niche. Int J Hematol 2017; 106(1):45–54. doi: https://doi.org/10.1007/s12185-017-2262-9.
13. Müller-Sieburg CE, Cho RH, Thoman M, Adkins B, Sieburg HB. Deterministic regulation of hematopoietic stem cell self-renewal and differentiation. Blood 2002; 100(4):1302–9.
14. Dykstra B, Kent D, Bowie M, McCaffrey L, Hamilton M, Lyons K, Lee SJ, Brinkman R, Eaves C. Long-term propagation of distinct hematopoietic differentiation programs in vivo. Cell Stem Cell 2007;1(2):218–29. doi: https://doi.org/10.1016/j.stem.2007.05.015.
15. Furutachi S, Matsumoto A, Nakayama KI, Gotoh Y. p57 controls adult neural stem cell quiescence and modulates the pace of lifelong neurogenesis. EMBO J 2013;32(7):970–81. doi: https://doi.org/10.1038/emboj.2013.50.
16. Kassem M, Bianco P. Skeletal stem cells in space and time. Cell 2015;160(1–2):17–9. doi: https://doi.org/10.1016/j.cell.2014.12.034.
17. Čamernik K, Mihelič A, Mihalič R, Haring G, Herman S, Marolt Presen D, et al. Comprehensive analysis of skeletal muscle- and bone-derived mesenchymal stem/stromal cells in patients with osteoarthritis and femoral neck fracture. Stem Cell Res Ther 2020;11(1):146. doi: https://doi.org/10.1186/s13287-020-01657-z.
18. Kawamura H, Nakatsuka R, Matsuoka Y, Sumide K, Fujioka T, Asano H, et al. TGF-β Signaling Accelerates Senescence of Human Bone-Derived CD271 and SSEA-4 Double-Positive Mesenchymal Stromal Cells. Stem Cell Reports 2018;10(3):920–932. doi: https://doi.org/10.1016/j.stemcr.2018.01.030.
19. Montagnani S, Rueger MA, Hosoda T, Nurzynska D. Adult stem cells in tissue maintenance and regeneration. Stem Cells Int 2016; 2016:7362879.
20. Blanpain C, Fuchs E. Plasticity of epithelial stem cells in tissue regeneration. Science 2014; 344:1242281.
21. Saghizadeh M, Kramerov AA, Svendsen CN, Ljubimov AV. Concise review: stem cells for corneal wound healing. Stem Cells 2017; 35:2105-2114.
22. Barbaro V, Orvieto A, Alvisi G, Bertolin M, Bonelli F, Liehr T, et al. Analysis and pharmacological modulation of senescence in human epithelial stem cells. J Cell Mol Med 2022;26(14):3977–3994
23. Notta F, Doulatov S, Laurenti E, Poeppl A, Jurisica I, Dick JE Isolation of single human hematopoietic stem cells capable of long-term multilineage engraftment. Science 2011; 333:218–221. doi: https://doi.org/10.1126/science.1201219.
24. Latchney SE, Calvi LM. The aging hematopoietic stem cell niche: Phenotypic and functional changes and mechanisms that contribute to hematopoietic aging. Semin Hematol 2017; 54:25–32. doi: https://doi.org/10.1053/j.seminhematol.2016.10.001
25. Pang WW, Schrier SL, Weissman IL. Age-associated changes in human hematopoietic stem cells. Semin Hematol 2017; 54:39–42. doi: 10.1053/j. seminhematol.2016.10.004.
26. Pietras EM, Warr MR, Passegue E. Cell cycle regulation in hematopoietic stem cells. J Cell Biol 2011; 195:709–720. doi: https://doi.org/10.1083/jcb.201102131.

27. Clapes T, Lefkopoulos S, Trompouki E. Stress and Non-Stress Roles of Inflammatory Signals during HSC Emergence and Maintenance. Front Immunol 2016; 7:487. doi: https://doi.org/10.3389/fimmu.2016.00487.
28. Gao X, Xu C, Asada N, Frenette PS. The hematopoietic stem cell niche: From embryo to adult. Development 2018; 145 doi: https://doi.org/10.1242/dev.139691.
29. Sacma M, Pospiech J, Bogeska R, de Back W, Mallm JP, Sakk V, et al. Haematopoietic stem cells in perisinusoidal niches are protected from ageing. Nat Cell Biol 2019; 21:1309–1320. doi: https://doi.org/10.1038/s41556-019-0418-y.
30. Cheng H, Zheng Z, Cheng T. New paradigms on hematopoietic stem cell differentiation. Protein Cell 2020; 11:34–44. doi: https://doi.org/10.1007/s13238-019-0633-0.
31. Laurenti E, Gottgens B. From haematopoietic stem cells to complex differentiation landscapes. Nature 2018; 553:418–426. doi: https://doi.org/10.1038/nature25022.
32. Zhang Y, Gao S, Xia J, Liu F. Hematopoietic Hierarchy—An Updated Roadmap. Trends Cell Biol 2018; 28:976–986. doi: https://doi.org/10.1016/j.tcb.2018.06.001.
33. Singh S, Jakubison B, Keller JR. Protection of hematopoietic stem cells from stress-induced exhaustion and aging. Curr Opin Hematol. 2020 Jul;27(4):225–231. doi: https://doi.org/10.1097/MOH.0000000000000586.
34. Rübe CE, Fricke A, Widmann TA, Fürst T, Madry H, Pfreundschuh M, Rübe C. Accumulation of DNA damage in hematopoietic stem and progenitor cells during human aging. PLoS One 2011;6(3):e17487. doi: https://doi.org/10.1371/journal.pone.0017487.
35. Mitchell E, Chapman MS, Williams N, et al. 2021. Clonal dynamics of haematopoiesis across the human lifespan. bioRxiv 2021.08.16.456475.
36. Challen GA & Goodell MA Clonal hematopoiesis: mechanisms driving dominance of stem cell clones. Blood 2020; 136: 1590–1598.
37. Jaiswal S & Ebert BL Clonal hematopoiesis in human aging and disease. Science 2019; 366: eaan4673.
38. Jeong M et al. Loss of Dnmt3a immortalizes hematopoietic stem cells in vivo. Cell Rep 2018; 23: 1–10.
39. Tovy A et al. Tissue-biased expansion of DNMT3A-mutant clones in a mosaic individual is associated with conserved epigenetic erosion. Cell Stem Cell 2020; 27: 326–335.e4.
40. Ahmad H, Jahn N, Jaiswal S. Clonal hematopoiesis and its impact in human health. Annu Rev Med 2023; 74: 249–260
41. He H et al. Aging-induced IL27Ra signaling impairs hematopoietic stem cells. Blood 2020; 136: 183–198
42. Valletta S et al. Micro-environmental sensing by bone marrow stroma identifies IL-6 and TGFbeta1 as regulators of hematopoietic ageing. Nat. Commun 2020; 11: 4075
43. Ho YH et al. Remodeling of bone marrow hematopoietic stem cell niches promotes myeloid cell expansion during premature or physiological aging. Cell Stem Cell 2019; 25: e6
44. Frisch BJ et al. Aged marrow macrophages expand platelet-biased hematopoietic stem cells via interleukin-1B. JCI Insight 2019; 5: e124213.
45. Pang WW, Price EA, Sahoo D, Beerman I, Maloney WJ, Rossi DJ, et al. Human bone marrow hematopoietic stem cells are increased in frequency and myeloid-biased with age. PNAS 2011; 108(50):20012–7. https://doi.org/10.1073/pnas.1116110108
46. Rundberg Nilsson A, Soneji S, Adolfsson S, Bryder D, Pronk CJ. Human and Murine Hematopoietic Stem Cell Aging Is Associated with Functional Impairments and Intrinsic Megakaryocytic/Erythroid Bias. PLoS One 2016;11(7):e0158369. doi: https://doi.org/10.1371/journal.pone.0158369.
47. Esplin BL, Shimazu T, Welner RS, Garrett KP, Nie L, Zhang Q, et al. Chronic exposure to a TLR ligand injures hematopoietic stem cells. J. Immunol 2011; 186:5367–5375. https://doi.org/10.4049/jimmunol.1003438
48. Mann M, Mehta A, de Boer CG, Kowalczyk MS, Lee K, Haldeman P, et al. Heterogeneous Responses of Hematopoietic Stem Cells to Inflammatory Stimuli Are Altered with Age. Cell Rep 2018; 25:2992–3005.e5. https://doi.org/10.1016/j.celrep.2018.11.056

49. Matatall KA, Jeong M, Chen S, Sun D, Chen F, Mo Q, et al. Chronic Infection Depletes Hematopoietic Stem Cells through Stress-Induced Terminal Differentiation. Cell Rep 2016; 17:2584–2595. https://doi.org/10.1016/j.celrep.2016.11.031
50. Abegunde SO, Buckstein R, Wells RA, and Rauh MJ. An inflammatory environment containing TNFα favors Tet2-mutant clonal hematopoiesis. Exp. Hematol 2018; 59:60–65. https://doi.org/10.1016/j.exphem.2017.11.002
51. Cai Z, Kotzin JJ, Ramdas B, Chen S, Nelanuthala S, Palam LR, et al. Inhibition of Inflammatory Signaling in Tet2 Mutant Preleukemic Cells Mitigates Stress-Induced Abnormalities and Clonal Hematopoiesis. Cell Stem Cell 2018; 23:833–849.e5. https://doi.org/10.1016/j.stem.2018.10.013
52. Meisel M, Hinterleitner R, Pacis A, Chen L, Earley ZM, Mayassi T, et al. Microbial signals drive pre-leukaemic myeloproliferation in a Tet2-deficient host. Nature 2018; 557:580–584. https://doi.org/10.1038/s41586-018-0125-z
53. Leoni C, Montagner S, Rinaldi A, Bertoni F, Polletti S, Balestrieri C and Monticelli S. Dnmt3a restrains mast cell inflammatory responses. Proc Natl Acad Sci USA 2017; 114:E1490–E1499. https://doi.org/10.1073/pnas.1616420114
54. King KY, Goodell MA. Inflammatory modulation of HSCs: viewing the HSC as a foundation for the immune response. Nat Rev Immunol 2011; 11:685–692. https://doi.org/10.1038/nri3062
55. Nagai Y, Garrett KP, Ohta S, Bahrun U, Kouro T, Akira S, et al. Toll-like receptors on hematopoietic progenitor cells stimulate innate immune system replenishment. Immunity 2006; 24:801–812. https://doi.org/10.1016/j.immuni.2006.04.008
56. Baldridge MT, King KY, Boles NC, Weksberg DC, Goodell MA. Quiescent hematopoietic stem cells are activated by IFN-gamma in response to chronic infection. Nature 2010; 465:793–797. https://doi.org/10.1038/nature09135
57. Mossadegh-Keller N, Sarrazin S, Kandalla PK, Espinosa L, Stanley ER, et al. M-CSF instructs myeloid lineage fate in single haematopoietic stem cells. Nature 2013; 497:239–243. https://doi.org/10.1038/nature12026
58. Hormaechea-Agulla D, Le DT, and King KY. Common Sources of Inflammation and Their Impact on Hematopoietic Stem Cell Biology. Curr Stem Cell Rep 2020; 6:96–107. https://doi.org/10.1007/s40778-020-00177-z
59. Shevyrev D, Tereshchenko V, Blinova E, Knauer N, Pashkina E, Sizikov A, Kozlov V Regulatory T Cells Fail to Suppress Fast Homeostatic Proliferation In Vitro. Life 2021; 11:245. doi: https://doi.org/10.3390/life11030245.
60. Min B. Spontaneous T Cell Proliferation: A Physiologic Process to Create and Maintain Homeostatic Balance and Diversity of the Immune System. Front Immunol 2018; 9:547. doi: https://doi.org/10.3389/fimmu.2018.00547.
61. Shevyrev D, Blinova EA, Kozlov VA. The influence of humoral factors of homeostatistic proliferation on t-regulatory cells in vitro. Bull Sib Med 2019; 18:286–293. doi: https://doi.org/10.20538/1682-0363-2019-1-286-293.
62. Lin Y, Kim J, Metter EJ, Nguyen H, Truong T, Lustig A, et all. Changes in blood lymphocyte numbers with age in vivo and their association with the levels of cytokines/cytokine receptors. Immun Ageing 2016; 13:24. doi: https://doi.org/10.1186/s12979-016-0079-7.
63. Thomas R, Wang W, Su DM. Contributions of Age-Related Thymic Involution to Immunosenescence and Inflammaging. Immun Ageing 2020; 17:2. doi: https://doi.org/10.1186/s12979-020-0173-8.
64. Britanova OV, Putintseva EV, Shugay M, Merzlyak EM, Turchaninova MA, Staroverov DB, et al. Age-related decrease in TCR repertoire diversity measured with deep and normalized sequence profiling. J Immunol 2014; 192:2689–2698. doi: 10. 4049/jimmunol.1302064.
65. Sun X, Nguyen T, Achour A, Ko A, Cifello J, Ling C, et al. Longitudinal analysis reveals age-related changes in the T cell receptor repertoire of human T cell subsets. J Clin Investig 2022; 132:e158122. doi: https://doi.org/10.1172/JCI158122.
66. Salminen A. Clinical perspectives on the age-related increase of immunosuppressive activity. J Mol Med 2022; 100:697–712. doi: https://doi.org/10.1007/s00109-022-02193-4.

67. Lee KA, Flores RR, Jang IH, Saathoff A, Robbins PD. Immune Senescence, Immunosenescence and Aging. Front Aging 2022; 3:900028. doi: https://doi.org/10.3389/fragi. 2022.900028.
68. Yousefzadeh MJ, Flores RR, Zhu Y, Schmiechen ZC, Brooks RW, Trussoni CE, et al. An aged immune system drives senescence and ageing of solid organs. Nature 2021; 594:100–105. doi: https://doi.org/10.1038/s41586-021-03547-7.
69. Li W. Phagocyte dysfunction, tissue aging and degeneration. Ageing Res Rev 2013; 12:1005–1012. doi: https://doi.org/10.1016/j.arr.2013.05.006. [
70. Uhl B, Vadlau Y, Zuchtriegel G, Nekolla K, Sharaf K, Gaertner F, et al. Aged neutrophils contribute to the first line of defense in the acute inflammatory response. Blood 2016; 128:2327–2337. doi: https://doi.org/10.1182/blood-2016-05-718999.
71. Tanaka N, Kato H, Tsuda H, Sato Y, Muramatsu T, Iguchi A, et al. Development of a High-Efficacy Reprogramming Method for Generating Human Induced Pluripotent Stem (iPS) Cells from Pathologic and Senescent Somatic Cells. Int J Mol Sci 2020; 21(18):6764. https://doi.org/https://doi.org/10.3390/ijms21186764
72. Vosough M, Ravaioli F, Zabulica M, Capri M, Garagnani P, Franceschi J, et al. Applying hydrodynamic pressure to efficiently generate induced pluripotent stem cells via reprogramming of centenarian skin fibroblast. PLoS One 2019; 14(4): e0215490 doi: https://doi.org/10.1371/journal.pone.0215490
73. Saini JS, Corneo JD, Miller RR, Kiehl TR, Wang N, Boles TA, et al. Nicotinamide ameliorates disease phenotypes in a human iPSC model of age-related macular degeneration. Cell Stem Cell 2017; 20(5): 635–647.e7 doi: https://doi.org/10.1016/j.stem.2016.12.01
74. Gong H, Cai NS, NYSCF Global Stem Cell Array Team. Stem cell-derived retinal pigment epithelium from patients with age-related macular degeneration exhibit reduced metabolism and matrix interactions. Stem Cells Transl Med 2020; 9(3): 364–376
75. Yasumoto T, Takamura Y, Tsuji M, Watanabe-Nakayama T, Imamura K, Inoue H, et al. High molecular weight amyloid β1-42 oligomers induce neurotoxicity via plasma membrane damage. FASEB J 2019; 33(8): 9220–9234. https://doi.org/10.1096/fj.201900604R
76. Kondo T, Yamamoto T, Okayama K, Narumi H, Inoue H. Metabolites of soil microorganisms modulate amyloid β production in Alzheimer's neurons. Sci Rep 2022; 2690 https://doi.org/10.1038/s41598-022-06513-z
77. Jarosz-Griffiths, Corbett NJ, Rowland HA, Fisher AC, Jones AC, Baron J. Proteolytic shedding of the prion protein via activation of metallopeptidase ADAM10 reduces cellular binding and toxicity of amyloid- β oligomers. J Biol Chem 2019; 294(17): 7085–7097.
78. Zhang J, Wilson GF, Soerens AG, Koonce CH, Yu J, Palecek SP, et al. Functional cardiomyocytes derived from human induced pluripotent stem cells. Circ Res 2009; 104(4): e30-e42 https://doi.org/10.1161/CIRCRESAHA.108.192237
79. Parikh SS, Blackwell DJ, Gomez-Hurtado M, Frisk L, Wang K, Kim CP et al. Thyroid and glucocorticoid hormones promote functional T-tubule development in human-induced pluripotent stem cell-derived cardiomyocytes. Circ Res 2017; 121(12): 1323–1330. https://doi.org/10.1161/CIRCRESAHA.117.311920
80. Ronaldson K, Bourchard SP, Ma K, Yeager T, Chen L, Song D, et al. Advanced maturation of human cardiac tissue grown from pluripotent stem cells. Nature 2018; 556 (7700): 239–243. https://doi.org/10.1038/s41586-018-0016-3
81. Theodoris CV, Zhou L, Liu Y, Zhang T, Nishino Y, Huang A, et al. Network-based screen in iPSC-derived cells reveals therapeutic candidate for heart valve disease. Science 2021; 371 (6530) eabd0724. https://doi.org/10.1126/science.abd0724
82. Gong L, Yan Q, Zhang Y, Fang X, Liu B, Guan X. Cancer cell reprogramming: a promising therapy converting malignancy to benignity. Cancer Commun 2019; 39(1): 48 https://doi.org/10.1186/s40880-019-0393-5
83. Hwang JW, Desterke C, Féraud S, Richard S, Ferlicot S, Verkatte V, et al. iPSC-derived embryoid bodies as model of c-met-mutated hereditary papillary renal cell carcinoma. Int J Mol Sci 2019; 20(19): 4867 https://doi.org/10.3390/ijms20194867
84. Hwang JW, Loisel-Duwattez J, Desterke C, Latsis T, Pagliaro S, Griscelli F, et al. A novel neuronal organoid model mimicking glioblastoma (GBM) features from induced pluripotent

stem cells (iPSC). Biochim Biophys Acta 2020; 1864(4): 129540 https://doi.org/10.1016/j.bbagen.2020.129540
85. Huang L, Desai R, Conrad DN, Leite NC, Akshinthala D, Lim CM, et al. Commitment and oncogene-induced plasticity of human stem cell-derived pancreatic acinar and ductal organoids. Cell Stem Cell 2021; 28(6): 1090–1104.e6 https://doi.org/10.1016/j.stem.2021.03.022
86. Dost A, Moye AL, Vedaie M, Tran LM, Fung E, Heinze D, et al. Organoids model transcriptional hallmarks of oncogenic KRAS activation in lung epithelial progenitor cells. Cell Stem Cell 2020; 27(4): 665–678.e8 https://doi.org/10.1016/j.stem.2020.07.022
87. Crespo M, Vilar E, Tsai SY, Chang K, Amin S, Srinivasan T, et al. Colonic organoids derived from human induced pluripotent stem cells for modeling colorectal cancer and drug testing. Nat Med 2017; 23(7): 878–884 https://doi.org/10.1038/nm.4355
88. Ito T, Kawai Y, Yasui Y, Iriguchi S, Minagawa A, Ishii T, et al. The therapeutic potential of multiclonal tumoricidal T cells derived from tumor infiltrating lymphocyte-derived iPS cells. Commun Biol 2021; 4(1): 694 https://doi.org/10.1038/s42003-021-02195-x
89. Zhu H, Blum RH, Bjordahl R, Gaidarova S, Rogers P, Lee TT, et al. Pluripotent stem cell-derived NK cells with high-affinity noncleavable CD16a mediate improved antitumor activity. Blood 2020; 135(6): 399–410 https://doi.org/10.1182/blood.2019000621
90. Mazza R, Maher J. Prospects for development of induced pluripotent stem cell-derived CAR-targeted immunotherapies. Arch Immunol Ther Exp 2021; 70(1): 2 https://doi.org/10.1007/s00005-021-00640-7
91. Park JH, Jain N, Chen A, McGuirk JP, Diaz M, Valamer B, et al. A phase I study of FT819, a first-of-kind, of-the-shelf, iPSC-derived TCR-less CD19 CAR T cell therapy for the treatment of relapsed refractory B-cell malignancies. Blood 2020; 136(suppl 1): 15–16 https://doi.org/10.1182/blood-2020-142423

Chapter 8
Cell-to-Cell Communication in Aging: Mechanisms, Impact and Therapeutic Prospects

Cristina Mas-Bargues, Jorge Sanz-Ros, Javier Huete-Acevedo, and Consuelo Borrás

Introduction

Aging is the outcome of the gradual accumulation of changes within an organism over time, resulting in a decline in biological function. It is a universal phenomenon that affects all living organisms and is characterized by a progressive reduction in both physical and mental capabilities, which is associated with an increased susceptibility to illness and mortality [1–3]. The biological process of aging is a complex phenomenon that occurs due to the accumulation of a variety of molecular and cellular damages over time. These damages lead to gradual changes in organs and systems that affect their function in the body [4]. Lopez-Otín et al. described the twelve hallmarks of aging, which represent common molecular and cellular damages that occur during aging in different organisms [5]. The Hallmarks of Aging include genomic instability, telomere shortening, epigenetic alterations, loss of proteostasis, disabled macroautophagy, abnormal regulation of nutrient sensing, mitochondrial dysfunction, cellular senescence, depletion of stem cells, altered intercellular communication, chronic inflammation, and dysbiosis. Investigating these hallmarks of aging is

C. Mas-Bargues · J. Sanz-Ros · J. Huete-Acevedo · C. Borrás (✉)
MiniAging Research Group, Department of Physiology, Faculty of Medicine, Centro de Investigación Biomédica en Red Fragilidad y Envejecimiento Saludable-Instituto de Salud Carlos III (CIBERFES-ISCIII), INCLIVA, University of Valencia, 46010 Valencia, Spain
e-mail: consuelo.borras@uv.es

C. Mas-Bargues
e-mail: cristina.mas@uv.es

J. Sanz-Ros
e-mail: jorge.m.sanz@uv.es

J. Huete-Acevedo
e-mail: javier.huete@uv.es

V. Bueno (ed.), *Cellular and Molecular Aspects of Ageing*,
https://doi.org/10.1007/978-3-031-55022-5_8

important because it helps us understand the complex aging process and identify therapeutic targets to treat age-related pathologies, thereby improving life expectancy and quality of life during aging.

Concept of Cell-to-Cell Communication

Cell-to-cell communication is a vital biological process necessary for the maintenance of homeostasis in multicellular organisms. This process depends on the coordination between cells to perform their functions effectively. It plays an important role in different tasks such as cell development, differentiation, and the ability to adapt to the environment [6–8].

Cell-to-cell communication generally involves the exchange of biochemical substances between cells, including ions, metabolites, hormones, cytokines, etc. This type of communication is commonly referred to as biochemical signaling. There are two types of biochemical signaling: direct communication (juxtacrine) and indirect communication (autocrine, paracrine, and endocrine). However, classically, these four different types of intercellular communication have been classified according to the distance between the sending and receiving cells [9]:

- Direct, contact-dependent, or juxtacrine communication occurs between neighboring cells near each other in which their membrane structures facilitate the exchange of molecules without these being secreted into the extracellular space. There are three direct biochemical communication pathways:
  - Communication mediated by cell adhesion molecules: selectins, integrins, cadherins, and immunoglobulins.
  - Communication mediated by gap junctions, which are formed by two hemichannels (oligomers of 6 intrinsic membrane proteins, called connexins) inserted where two cells are contiguous, and precisely aligned so that the lumen of one is continuous with that of the other.
  - Communication mediated by tunneling nanotubes, which are thin projections of the plasma membrane.
- Indirect communication is characterized by the secretion of signaling molecules, which are transported by flow or diffusion. These molecules are released into the extracellular environment and can either diffuse passively across the cell membrane (in the case of hydrophobic small molecules) or be internalized into the cytoplasm of the target cell through endocytosis (for hydrophilic molecules). Another type of biochemical indirect communication involves the secretion of extracellular vesicles by the signaling cell, which contains the signaling molecules. Based on the distance between the secretory cell and the target cell, biochemical indirect communication can be classified as:
  - Paracrine: refers to signaling molecules that act on cells near the sending cell, diffusing through space and the extracellular matrix. To achieve this local

action, ligands are usually degraded by enzymes or eliminated by nearby cells. This communication only affects cells in the immediate neighborhood of the secretory cell.

- Autocrine: this is the signaling that occurs when the sender and the receptor cell are the same cell so that the cell secretes factors into the extracellular space and when it encounters its own receptors, they activate intracellular signaling cascades. Thus, signals are directed to the cells that secrete them.
- Endocrine: when the sending and receiving cells are located at distant points in the body. Molecules are released from endocrine cells and generally have a longer-lasting effect on target cells than the other types of communication. Hormones are the characteristic molecules of this type of communication, and according to the classical concept, it is the endocrine glands (thyroid, pancreas, hypothalamus, etc.) that are responsible for releasing these molecules. More recently, however, it has been discovered that most tissues release to a greater or lesser extent factors that serve as signals for distant tissues [10–12].

In addition to biochemical communication, there is also mechanical communication. In this type of cell-to-cell communication, cells interact with their environment by applying mechanical forces to the extracellular matrix. These signals play a crucial role in several processes, including the reorganization of the cytoskeleton and cell migration [13], the remodeling of the extracellular matrix [14], epigenetic modifications [12], and the recruitment of immune cells to the tissue [15], among other functions.

Importance of Understanding How Cell-to-Cell Communication Influences Aging

As mentioned earlier, this alteration in cell-to-cell communication is classified as a hallmark of aging. Specifically, it is defined as an integrative hallmark that arises when the accumulated damage from the primary hallmarks and antagonist hallmarks cannot be compensated by tissue homeostatic mechanisms. In other words, this integrative hallmark directly impacts tissue homeostasis and its function, making it responsible for the alteration of the phenotype during aging [5]. During aging, intercellular communication is disrupted at all levels: autocrine, paracrine, endocrine, and neuroendocrine. This dysregulation tends to lead to a decrease in immunosurveillance against pathogens and premalignant cells and an increase in inflammatory reactions [16]. A significant change in intercellular communication that is closely associated with aging is the emergence of a chronic and low-grade systemic inflammatory process known as inflammaging. This inflammaging condition, which will be explained later, contributes to the development of age-related diseases such as obesity, type 2 diabetes, atherosclerosis, and Alzheimer's disease [17, 18].

Scope and Objectives of This Review

In this review chapter, we try to elucidate the complex processes of cell-to-cell communication in the context of aging. It includes the molecular and cellular mechanisms involved leading to the spread of cellular senescence and the inherent chronic inflammation. Therefore, this review aims to provide an overview of the mechanisms controlling intercellular communication in aging and to describe how changes in intercellular communication affect the aging process and the development of age-related diseases. Finally, this review aims to explain the importance of understanding these mechanisms in developing potential therapeutic interventions aimed at restoring or modulating cell-to-cell communication to mitigate the effects of aging and age-related diseases.

Mechanisms of Cell-to-Cell Communication in Aging

Intercellular Signaling Pathways

For a multicellular organism to function properly, its constituent cells must communicate with each other and influence each other's behavior; even in single-celled organisms such as bacteria or yeast, communication between neighboring cells is necessary for the colony to thrive. There are thousands of molecules that mediate signaling between cells; proteins, peptides, amino acids, nucleotides, steroids, retinoids, fatty acids, and even dissolved gases such as nitric oxide or carbon monoxide. These molecules are released into the extracellular space by exocytosis, by diffusion across the membrane, through cell–cell junctions, or by using extracellular vesicles as a vehicle [19].

Target cells respond to these extracellular signals via receptors, which can be membrane, intracellular, or intranuclear, and bind specifically to the signaling molecule, and then generate a response in the cell via an intracellular signaling cascade that modifies the cell's behavior. A large number of signaling pathways have been described and all are based on this signaling molecule-receptor pairing.

Intercellular Communication Alterations During Aging

In addition to cell-intrinsic alterations, aging causes negative changes in cell–cell communication (Fig. 8.1). During aging, the extracellular environment suffers several changes, altering the concentration of different factors such as hormones, cytokines, and growth factors. For example, there is dysregulation of neuroendocrine signaling (renin-angiotensin axis, insulin-like growth factor 1 (IGF1) pathway) during aging. The research findings indicate that genetic manipulations that reduce signaling within

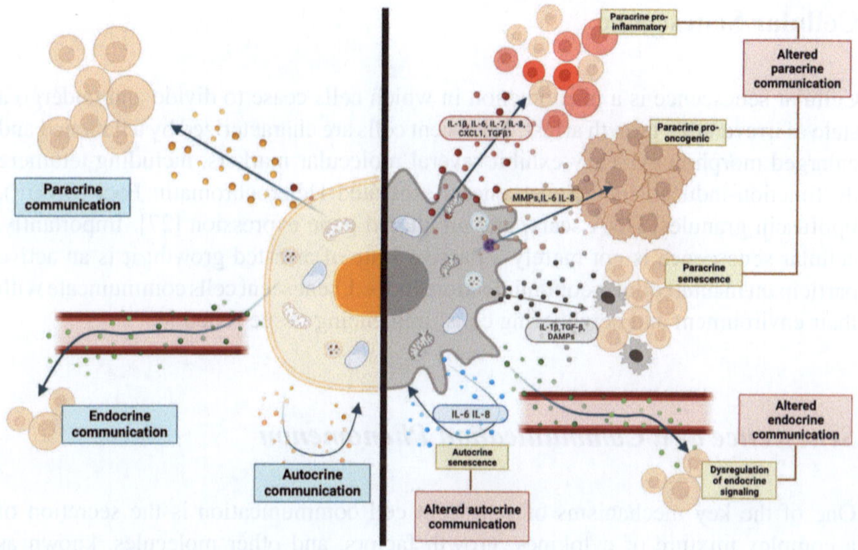

Fig. 8.1 Alterations in cell-to-cell communication with aging

the IGF1 pathway consistently led to extended lifespans in worms, flies, and mice. This pathway plays a significant role in mediating the positive effects of dietary restriction on longevity in worms and flies [20].

As individuals age, pro-inflammatory reactions increase, immune surveillance decreases, and overall, there is a change in the composition of the extracellular milieu, affecting tissue function [21]. The most studied alteration of intercellular communication in aging is the so-called "inflammaging", which refers to the phenotype that accompanies individuals during aging, characterized by a systemic pro-inflammatory state [22]. This systemic inflammation is closely related to diseases that accompany aging, such as type 2 diabetes or atherosclerosis, and also contributes to the loss of regenerative capacity in some tissues such as the skin [17]. This particular scenario will be discussed further in an extended section.

In addition to the systemic inflammation that accompanies aging, there is evidence showing that aging-related changes occurring in one tissue are capable of producing similar changes in other distant tissues leading to their deterioration. For example, an aged microenvironment contributes to CD4 T-lymphocyte dysfunction [23]; similarly, age-related loss of kidney function leads to an increased risk of cardiovascular disease [24]. On the other hand, tissue-specific interventions on various aging processes can delay aging in distant tissues, for example, induction of telomerase expression in the gut alone has been shown to counteract the aging of the whole organism [25], and modulation of mitochondrial activity in key tissues can improve longevity and delay systemic aging [26].

Cellular Senescence

Cellular senescence is a phenomenon in which cells cease to divide and undergo a state of irreversible growth arrest. Senescent cells are characterized by a flattened and enlarged morphology. They exhibit several molecular markers, including telomere dysfunction-induced foci, Senescence-Associated Heterochromatin Foci (SAHF), lipofuscin granules, DNA scars, and an altered gene expression [27]. Importantly, cellular senescence is not merely a passive state of arrested growth; it is an active participant in intercellular communication. Indeed, senescent cells communicate with their environment and neighboring cells, influencing tissue function.

Senescence as a Communication Phenomenon

One of the key mechanisms of senescent cell communication is the secretion of a complex mixture of cytokines, growth factors, and other molecules, known as the Senescence-Associated Secretory Phenotype (SASP) [16]. The SASP serves a dual purpose. It can attract immune cells to remove senescent cells, but it can also cause inflammation and damage to surrounding tissues if not properly regulated. For example, in the context of chronic inflammation associated with aging (inflammaging), the SASP can lead to age-related diseases such as atherosclerosis, neurodegenerative disorders, and cancer.

SASP serves for multiple purposes (Fig. 8.2), which include:

- Facilitating the clearance of senescent cells by the immune system: The immune system, including natural killer (NK) cells and macrophages, plays a crucial role in responding to senescent cell signals [28]. SASP components act as chemo-attractive signals for immune cells to recognize and eliminate senescent cells. This process is vital for maintaining tissue integrity and preventing the accumulation of damaged cells.
- Direct cell-to-cell interactions: Senescent cells can communicate with neighboring cells through direct interactions. They may induce cell-cycle arrest in adjacent proliferating cells via paracrine signaling. This arrest prevents the proliferation of damaged or stressed cells and has implications for tissue homeostasis. In the context of cancer, senescence-associated cell interactions are essential for preventing the proliferation of malignant cells [29]. The senescence of cancer cells can halt tumor growth and potentially aid in their removal by the immune system.
- Promoting tissue repair and regeneration: Senescence-associated cell interactions play a significant role in tissue regeneration. Cell populations involved in wound healing such as vascular, immune, and mesenchymal cells adopt senescence features including a related secretome rich in Platelet-derived growth factor alpha polypeptide a (PDGF-AA) [30]. These senescent cells contribute to wound

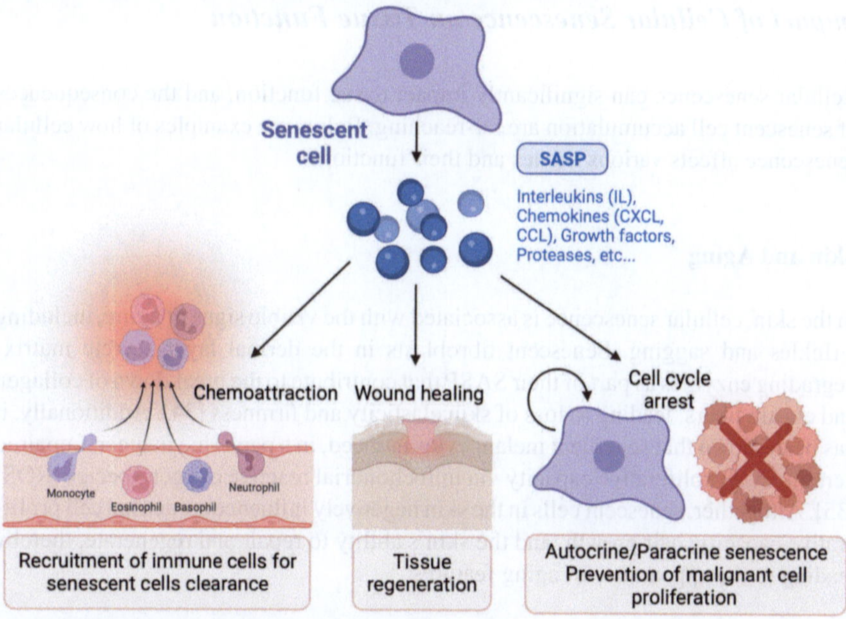

Fig. 8.2 Beneficial effects of SASP in adults

healing and tissue repair by promoting a microenvironment conducive to tissue remodeling and regeneration [31].

- Contributing to chronic inflammation when SASP is sustained: Immune cells are involved in the clearance of senescent cells. However, the aging of the immune system is characterized by reduced efficiency of the immune system, leading to an accumulation of senescent cells. This accumulation can result in a chronic inflammatory state, contributing to aging-related diseases [32]. An example of this is the role of senescent cells in osteoarthritis, where the immune response to these cells in joint tissues contributes to cartilage degradation.

Senescence is a communication phenomenon within biological systems, facilitated by the SASP and senescence-associated cell interactions. These mechanisms play critical roles in various physiological and pathological processes, including aging, tissue repair, and cancer suppression [33]. Understanding senescence as a communication process is essential for developing interventions to target age-related diseases and cancer while preserving tissue homeostasis.

Impact of Cellular Senescence on Tissue Function

Cellular senescence can significantly impact tissue function, and the consequences of senescent cell accumulation are far-reaching. Below are examples of how cellular senescence affects various tissues and their functions.

Skin and Aging

In the skin, cellular senescence is associated with the visible signs of aging, including wrinkles and sagging. Senescent fibroblasts in the dermal layer secrete matrix-degrading enzymes as part of their SASP that contribute to the breakdown of collagen and elastin fibers, leading to loss of skin elasticity and firmness [34]. Additionally, it has been shown that senescent melanocytes induced, in a paracrine manner, impaired keratinocyte proliferative capacity via mitochondrial reactive oxygen species (ROS) [35]. Altogether, senescent cells in the skin negatively influence epidermal cell proliferative capacity, hair growth, and the skin's ability to repair and regenerate, thereby leading to the appearance of aging features.

Brain and Neurodegeneration

The senescence of neural cells has been linked to neurodegenerative diseases such as Alzheimer's and Parkinson's disease. Transcriptomic analyses indicate a strong association between the presence of Tau-containing neurofibrillary tangles (NFTs) and neuronal senescence in the brain, which contributes to neurodegeneration [36]. Senescent astrocytes and microglia can contribute to chronic inflammation in the brain, leading to neuronal damage [37]. This communication can disrupt neuronal function, impair cognitive abilities, and promote Alzheimer's disease progression. Similarly, chronic inflammation by senescent cells in the midbrain triggers neuroinflammation resulting in the immune cell-mediated killing of midbrain dopaminergic neurons in Parkinson's disease [38].

Cardiovascular System and Atherosclerosis

In the cardiovascular system, senescent vascular endothelial cells and vascular smooth muscle cells are implicated in the development of atherosclerosis. Senescent vascular endothelial cells participate in atherosclerosis by regulating the levels of endothelin-1 (ET-1), nitric oxide (NO), angiotensin II, and monocyte chemoattractant protein-1 (MCP-1) [39–41]. These factors alter blood vessel integrity, leading to inflammation and plaque formation. Senescent vascular smooth muscle cells mediate plaque instability and vascular calcification via regulating the expression level of bone morphogenetic protein-2 (BMP-2), osteopontin (OPN), runt-related transcription factor (Runx-2) and proinflammatory cytokines such as interleukins and matrix metalloproteinases (MMPs) [42–45], which contribute to the stiffening of blood vessels and narrowing of arteries, increasing the risk of heart disease and stroke. Moreover, senescent macrophages impair cholesterol efflux and promote the development of senescent-related cardiovascular diseases [46].

Muscle and Sarcopenia

Sarcopenia, the age-related loss of muscle mass and strength, is influenced by cellular senescence. Muscle cells, also known as myocytes, when senescent, exhibit reduced regenerative capacity and secrete inflammatory molecules that impair muscle function [47]. In fact, a study sought to determine whether the implantation of senescent cells adjacent to skeletal muscle would affect healthy surrounding muscle cells via the "bystander effect" and senescence-related signaling [48]. Indeed, senescent cell transplantation adjacent to skeletal muscle induced an increase in senescent markers in the healthy muscle, as well as muscle-fiber thinning, indicative of sarcopenia [49]. This senescent-like communication leads to muscle atrophy and weakness, contributing to frailty in the elderly.

Bones and Osteoporosis

There is growing evidence to suggest that cellular senescence is a major player in the pathogenesis of osteoporosis, a prevalent age-related bone disorder characterized by deterioration of bone microarchitecture and low bone mineral density [49]. Senescent osteocyte accumulation and their secretory phenotype have been detected in the aged bone environment of mice and humans [50, 51]. Moreover, osteocyte senescence stimulates receptor activators of nuclear factor kappa-B (NFκB) ligand production, leading to increased osteoclast formation and activity, which accelerates the development of osteoporosis [52, 53].

Cellular senescence is not a passive aging process but a dynamic communication phenomenon that profoundly influences tissue function. Senescent cells communicate through mechanisms such as the SASP, paracrine signaling, and interactions with the immune system. While senescence can be a protective mechanism in response to injury or infection, the accumulation of senescent cells can contribute to various age-related diseases.

Immunosenescence and Inflammaging

Aging is a complex biological process characterized by a gradual decline in the body's functional capacity and an increased susceptibility to various diseases. The immune system aging is characterized by the paradox of immunosenescence (insufficiency) and inflammaging (over-reaction), which incorporates two sides of the same coin, resulting in immune disorder [54, 55]. Immunosenescence refers to a disruption in the structural architecture of immune organs and dysfunction in immune responses, resulting from both aged innate and adaptive immunity [56, 57]. Inflammaging, described as a chronic, sterile, systemic inflammatory condition associated with advanced age, is mainly attributed to the SASP and age-related autoimmune predisposition [17, 58]. This section focuses on these phenomena, exploring how they affect immune cells' communication their response to challenges, and ultimately the associated health consequences.

Immunosenescence: The Aging Immune System

Immunosenescence refers to the gradual deterioration of the immune system with advancing age [56, 57]. This process impacts various components of the immune system and compromises the body's ability to defend against infections and maintain tissue homeostasis. Several characteristic changes of immunosenescence are typically observed such as thymic involution, hematopoietic stem cell (HSC) dysfunction, disrupted naïve/memory ratio in T and B cells, accumulation of senescent cells, inflammaging, impaired new antigen response, mitochondrial dysfunction, genomic instability, and stress responses [59, 60] (Fig. 8.3).

Immunosenescence is not due to a lack of immune cells, but to a reduced immune repertoire diversity, attributed to insufficient production of naïve immune cells and amplified oligo-clonal expansion of memory immune cells. Immunosenescence is therefore linked to the thymus, a primary organ for T-cell development, that undergoes structural changes with age. As the thymus atrophies (thymus involution), the epithelial spaces gradually disappear, and the perivascular space fills the elderly thymus, leading to a decreased supply of naïve T cells [61, 62], which play a crucial role in recognizing new pathogens. As a result, older individuals have fewer T cells leading to a restricted T-cell receptor (TCR) repertoire. Indeed, elders are not capable of responding to novel threats, compromising the adaptive immune response.

Findings in experimental mouse models have described the main effects of T cell immunosenescence. $CD4^+$ T cells show reduced receptor signaling intensity [63]; defective activation, differentiation, and expansion after stimulation [64]; diminished production of interleukin-2 (IL-2) [65]; and impaired ability to provide cognate help to B cells after immunization, which is essential for antibody production [66, 67]. $CD8^+$ T cells display a declined responsiveness to newly encountered antigens (as a consequence of the reduced TCR repertoire) [68, 69]; increased clonal expansion

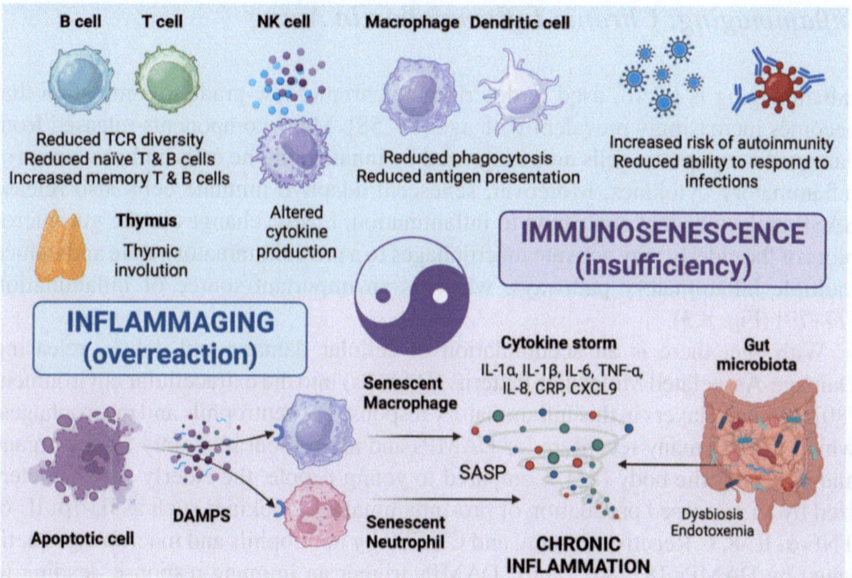

Fig. 8.3 The paradox of the immune system with aging: immunosenescence and inflammaging

of individual clones [70]; and impaired generation of memory T cells [71]. Taken together, these changes limit the immune system's ability to respond effectively to infections. As a result, older individuals may struggle to mount an effective immune response against infections or vaccinations.

Although immunosenescence is mostly related to failures in adaptive immunity (B and T cells), significant evidence indicates that aging also impacts various functions of the innate immune system [72]. With aging, neutrophils lose their ability to migrate and subsequently clear infections [73]. Defects in monocyte and macrophage function with age are known to be mediated by the dysregulation of cellular signaling [74]. Similarly, antigen-presenting cells (APCs) defective functionality during aging has been attributed to a decrease in Phosphoinositide 3-kinase (PI3K) activity, which is implicated in both APCs migration and Toll-like receptor (TLR) signaling [75]. Indeed, dendritic cells display a reduced antigen presentation capacity with aging, which affects the activation of T cells and the initiation of adaptive immune responses.

Paradoxically, as the immune system weakens in some aspects, it may become more prone to mounting inappropriate responses, including autoimmunity [76]. This means that aging individuals may face an increased risk of their immune system mistakenly attacking their own tissues. Overall, the aging process significantly alters the landscape of immune cell signaling and response. These changes result in a reduced ability to respond effectively to infections and an increased vulnerability to age-related diseases.

Inflammaging: Chronic Inflammation in Aging

Inflammaging is a term used to describe the chronic, low-grade inflammation that becomes increasingly prevalent with age [17, 58]. Many components released from damaged or apoptotic cells are recognized by innate immune cells and produce pro-inflammatory cytokines. Moreover, senescent adaptive immune cells also release SASP molecules that contribute to inflammation. Lastly, changes in the gut microbiota of the elderly may activate macrophages to a pro-inflammatory state and induce multiple inflammatory pathways, which is an important source of inflammation [77–79] (Fig. 8.3).

With age, there is an accumulation of cellular damage and debris, releasing Damage-Associated Molecular Patterns (DAMPs) into the extracellular environment [80]. The key players in this inflammatory response are neutrophils and macrophages, which express many receptors for DAMPs and are present in mainly all the organs and tissues of the body [81]. Compared to young people, the elderly are characterized by an increased production of pro-inflammatory cytokines such as IL-1β, IL-6, TNF-α, IL-8, C Reactive Protein, and CXCL9 by neutrophils and macrophages activated by DAMPs [82–84]. Thus, DAMPs trigger an immune response, leading to the production of pro-inflammatory mediators and the chronic inflammatory state associated with inflammaging.

Senescent immune cells exhibit a distinctive SASP that secretes pro-inflammatory cytokines. For instance, aged macrophages display a noteworthy increase in SASP pro-inflammatory components, such as TNF-α, IL-6, and IL-1β, as well as decreased levels of the anti-inflammatory cytokine IL-10 [85]. Memory B cells are more prevalent in older adults and can produce various pro-inflammatory cytokines and chemokines such as IL-1α, IL-1β, IL-6, and TNF-α, suggesting their potential involvement in inflammatory disorders during inflammaging [86]. To summarize, as the body ages, immune cells exhibit senescent characteristics, which manifest as an alteration of the cell-to-cell communication network, fostering a pro-inflammatory environment and leading to an impaired clearing function of the immune system.

The gut microbiome, consisting of trillions of microorganisms, plays a crucial role in overall health. With age, the composition of the gut microbiome undergoes significant changes, often referred to as dysbiosis. Specifically, the main age-associated changes in gut microbiota are a decrease in bacteria biodiversity and an increase in pathobionts (potentially harmful bacteria), which can result in a less stable microbial community [87]. Dysbiosis can lead to increased gut permeability, allowing bacterial components to enter the bloodstream. This process, often referred to as endotoxemia, triggers an inflammatory response and further propagates inflammaging.

Immunosenescence and inflammaging are critical components of the aging process, affecting how the immune system functions and contributing to chronic inflammation. These processes result in increased susceptibility to infections, a higher risk of age-related diseases, and tissue dysfunction.

Cell Communication and Longevity

Genetic and Dietary Influences

Longevity, the extension of an organism's lifespan, is a complex interplay of genetic and environmental factors. Among these factors, cell communication plays a pivotal role in influencing the aging process. Herein we explore the relationship between cell communication, genetics, dietary influences, and their collective impact on extending lifespan.

Genetics play a fundamental role in determining an individual's predisposition to age-related diseases and overall lifespan. It is estimated that about 25% of the variation in human life span is determined by genetics, but which genes, and how they exactly contribute to longevity, is still not well understood [88–91]. What is better understood is how genetic variations can affect cellular communication.

With age, genetic mutations occur and lead to several damages [92]. When these mutations affect DNA repair genes, it further contributes to genomic instability. In fact, many mutations in DNA repair genes cause replicative senescence and premature aging [93]. Neurological dysfunction is found in individuals and mouse models with genetic errors in DNA repair genes [94, 95]. Thus, individuals with intact DNA repair mechanisms, i.e., without mutations, have a more efficient DNA damage response (DDR) and can reduce the accumulation of DNA damage. This in turn can positively influence cell signaling pathways, thereby hampering cellular senescence and inflammation [96–98].

Telomeres, the protective caps at the ends of chromosomes, shorten with age [99]. Telomeres shorten as a result of cellular replication, leading to a permanent cell cycle arrest, also known as replicative senescence [100]. Additional mutational mechanisms can accelerate telomere shortening [101]. Thus, genetic variations can influence how quickly this shortening occurs, which in turn affects cell communication related to cellular aging and senescence.

Genetic variations can also impact the efficiency of the immune system [102, 103], leading to variations in the production of pro-inflammatory cytokines and cell communication between immune cells [104, 105]. As an example, genetic variability in the human immune system may affect susceptibility to, and severity of infection by severe acute respiratory syndrome coronavirus 2 (SARS-CoV-2), the virus responsible for the coronavirus disease (COVID-19). Genetic variations explain, at least in part, the differences in the strength of immune responses. Certain immune system genes code for proteins called human leukocyte antigens (HLA), which vary tremendously. These variations influence how well the immune system recognizes a given pathogen or other foreign entity. An HLA type that leads to poor recognition of SARS-CoV-2 could cause a person to be more vulnerable to the virus [106, 107].

Despite the genetic influences, lifestyle, and dietary patterns can also influence aging and longevity [108, 109]. Several studies have reported that the Mediterranean diet, characterized by a decreased consumption of saturated animal fat and red meat and a higher intake of fruits and vegetables, along with maintaining a healthy weight and reducing salt intake, together with an active lifestyle, such as engaging in physical and social activities and reducing smoking and alcohol consumption, can significantly modulate cell signaling pathways and extend lifespan [110–112].

Nutrient sensing pathways become deregulated with age and lose effectiveness with age. These pathways are a link between diet and aging [113]. Alterations in nutrient sensing pathways have gained increasing attention in the field of aging because they can be modulated with dietary interventions, such as high protein, low carbohydrate diets, or ketogenic diets [114]. Some of the nutrient-sensing pathways impaired by aging are the IGF1/PI3K/AKT/mTOR and AMPK/Sirtuin/PGC1α pathways [115–118]. These pathways are central to cell communication and metabolism, and their regulation can influence longevity. Indeed, it has been demonstrated that inhibition of mTOR in *S. cerevisiae* [119], *C. elegans* [120], *D. melanogaster* [121], and *M. musculus* [122] extends the life span in all four models. Interestingly, regulation of mTOR mitigates stem cell senescence [123, 124]. Similarly, overexpression of sirtuins promotes longevity in *S. cerevisiae* [125], *C. elegans* [126], *D. melanogaster* [127], and *M. musculus* [128]. Importantly, Sirt1, the best-studied member of the sirtuin family in mammals, is known to delay cellular senescence, mainly through the prevention of telomere attrition and the promotion of DNA damage repair [129].

Antioxidants in the diet, such as vitamins C and E, can reduce oxidative stress and the resultant cell damage (oxidative stress-induced premature senescence) which has been proven in in vitro studies [130, 131]. Both vitamin C and E inhibit cellular senescence associated with the PI3K/AKT/mTOR and the p53/p16/pRb pathways but also up-regulates the expression of other longevity factors such as FoxO3a [132, 133].

Effective cellular communication is crucial for various cellular functions, including response to environmental changes, immune system regulation, and tissue repair. Genetic factors may determine the baseline efficiency of these processes, while dietary factors can provide the necessary components for cells to function optimally, reducing the negative impact of aging on cellular communication. However, a holistic approach that combines genetics, diet, and lifestyle choices is essential for maintaining healthy cellular communication throughout the aging process.

Caloric Restriction and Hormesis

Caloric Restriction (CR), a dietary regimen that reduces calorie intake without causing malnutrition, is known to have hormetic effects on cell communication and aging [134–138]. Hormesis is a phenomenon in which a low dose of a stressor, such as caloric restriction, can have a beneficial, adaptive effect on cells.

Caloric restriction induces mild cellular stress, which activates cellular stress response pathways like the c-Jun N-terminal kinase pathway, the p38 mitogen-activated kinase regulation, and the Forkhead box transcription factor's signaling [139]. This, in turn, enhances the cellular antioxidant defense mechanisms and reduces oxidative stress. Findings from a 2-year CR clinical trial (CALERIE) in healthy non-obese humans provide new evidence of the persistent metabolic slowing down accompanied by reduced oxidative stress markers (NCT00427193 and NCT02695511) [140].

Autophagy is a housekeeping process that maintains cellular homeostasis under both normal physiologic and pathologic conditions. It is responsible for the recycling of metabolic macromolecules and damaged organelles through lysosomal degradation processes. Under stress conditions, such as nutrient deficiency, autophagy is activated to maintain proper cell function and promote cell survival [141]. Recent advances in CR research have suggested that the activation of autophagy is linked to the observed beneficial anti-aging effects. Evidence showed that CR induced a robust autophagy response in various metabolic tissues and that the inhibition of autophagy attenuated the anti-aging effects of CR, including oxidative damage [142–144].

Several studies conducted on animal models support the observation that CR can reduce inflammation [145]. CR reduces chronic inflammation by downregulating pro-inflammatory cell communication pathways [146]. It dampens the activation of NF-κB, a key player in the production of pro-inflammatory cytokines [147]. CR has proven to reduce the plasma concentration of cytokines, TNF-α and ICAM-1 and to induce cortisol release, which suppresses the systemic inflammatory [148] response.

As one of the main inducers of senescence is cellular damage accumulation, it is highly plausible that CR reduces the amount of senescent cells by preventing damage from occurring. Indeed, CR can protect against cellular deterioration in at least two major ways: interfering with the source of damage, for example via decreasing oxidative stress and inflammation, or repairing/eliminating already present damage by increasing autophagy [149]. Moreover, in obese mice models, implementation of 30% CR for 2 months notably decreased the levels of adipose tissue cytokines and chemokines, including IL-6, IL-2, IL-1Rα, MCP-1, and CXCL16, which are considered major components of SASP [150]. In hepatic tissue, even mild CR notably suppressed proinflammatory and lipogenic gene expression of molecules such as MCP-1, SREBPs, and peroxisome proliferator-activated receptor (PPAR)-γ [151]. Thus, CR reduces the SASP, and by doing so, it modulates cell-to-cell communication, preventing the spread of inflammation and promoting healthy tissue function.

Cell communication is a central player in the complex network of factors that influence longevity. Genetic variations can determine an individual's susceptibility to age-related diseases, while dietary interventions can modulate cell signaling pathways, reduce inflammation, and promote adaptive responses. The hormetic effects of caloric restriction on cell communication provide valuable insights into the potential for dietary strategies to extend lifespan and improve the quality of life in aging individuals.

Therapeutic Approaches

Targeting Cell Communication for Aging Interventions

As we have discussed, cell communication is a fundamental aspect of aging, and it plays a crucial role in the onset and progression of age-related diseases. As our understanding of these mechanisms grows, there is an interest in developing therapeutic approaches that specifically target intercellular communication to mitigate the effects of aging. Here we will explore the potential strategies for targeting cell communication and their implications for aging and age-related diseases.

Parabiosis

One of the first strategies developed to address the deterioration of intercellular communication during aging is parabiosis, which involves anatomically joining the vessels of two animals so that they share the circulatory system. Parabiosis experiments conducted in the past decade, based on the earlier work of McCay and colleagues [152], demonstrated the importance of intercellular communication and the microenvironment in aging [153–155]. In these studies, two syngeneic mice of different ages, one young and one old, were joined together in a way that did not lead to immune rejection. After a period of pairing, there was evidence of an improvement in several aging-associated parameters in the old mice that had been paired with young mice. This suggested the presence of factors in the plasma of the young mice that could be beneficial for various functions impaired in aging. In parallel, young mice subjected to this procedure showed accelerated aging (Fig. 8.4).

Circulating Anti-aging Factors

Parabiosis experiments lead to the conclusion that soluble factors present in the blood of young mice effectively restore renewal and repair capacity in old mice, which has precipitated the search for blood-borne anti-aging factors. Single-cell transcriptomic studies of tissues from mice subjected to heterochronic parabiosis have confirmed the capacity of young blood to rejuvenate multiple tissues and to restore age-associated changes in general gene expression [156, 157]. Researchers have identified some specific factors as potential candidates to be implemented for aging-related dysfunction. For instance, a protein called tissue inhibitor of metalloproteinases 2 (TIMP2) enriched in human umbilical cord plasma, young mouse plasma, and young mouse hippocampi, increases synaptic plasticity and cognitive function in aged mice. The anti-inflammatory interleukin IL-37, whose expression declines with age in monocytes, can improve endurance to exercise and ameliorate whole-body metabolism in old mice [158]. Growth differentiation factor 11 (GDF11) was one of the first candidates after parabiosis experiments, it has a rejuvenating effect on specific tissues like

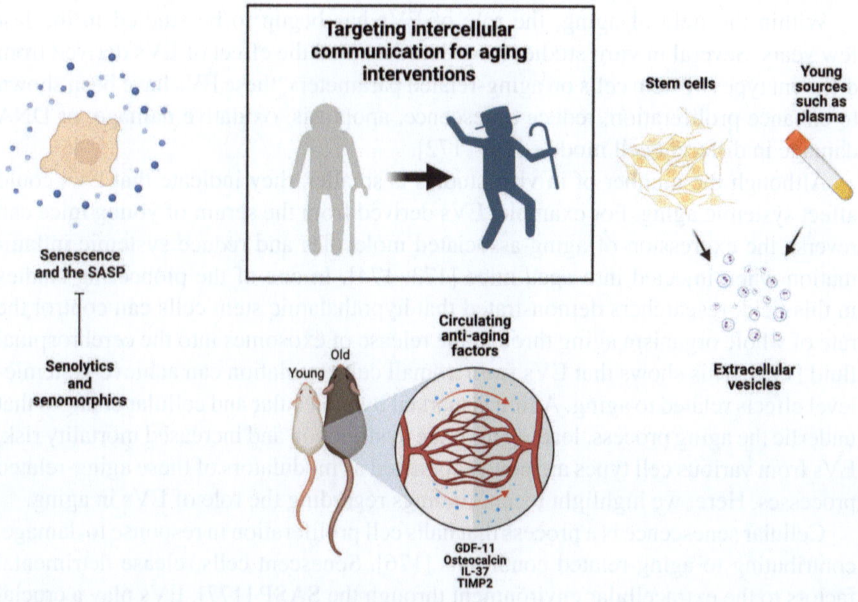

Fig. 8.4 Therapeutic approaches targeting aging

muscle, the brain, and the endocrine pancreas, yet it can hinder the function and repair of other tissues due to its pro-fibrotic side effects [159]. Additionally, osteocalcin (an osteoblast-derived hormone) levels decline with age, having an important role in cognitive function and responses to exercise, therefore is also proposed as a potential anti-aging factor [160].

Extracellular Vesicles

Another strategy on the rise is the use of extracellular vesicles (EVs) as therapeutics for aging and age-related diseases. EVs serve as vehicles for the transfer of different molecules between cells (proteins, lipids, DNA, RNA, microRNAs, etc.) and are involved in numerous physiological and pathological processes such as elimination of unwanted proteins, antigen presentation, genetic exchange or inflammation [161, 162]. The most studied types of EVs are exosomes and microvesicles (MVs). Exosomes are usually smaller and they are formed inside the cell through multivesicular bodies. On the other hand, MVs are larger, up to 1000 nm, although smaller MVs can also be found, MVs are formed directly from the plasma membrane [163].

EVs, particularly those coming from stem cells, have been shown to improve the regenerative capacity of different tissues. Stem cell studies have shown that they have limited integration potential [164–166], suggesting that the main effects of these cells are mediated by paracrine mechanisms [167]. Thus, its secretome, including EVs, has positioned as a potential therapy for tissue repair and regeneration [168].

Within the field of aging, the role of EVs has begun to be studied in the last few years. Several in vitro studies have demonstrated the effect of EVs derived from different types of stem cells on aging-related parameters, these EVs have been shown to enhance proliferation, reduce senescence, apoptosis, oxidative damage, or DNA damage in different cell models [169–172].

Although the number of in vivo studies is smaller, they indicate that EVs could affect systemic aging. For example, EVs derived from the serum of young mice can reverse the expression of aging-associated molecules and reduce systemic inflammation when injected into aged mice [173, 174]. In one of the pioneering studies in this field, researchers demonstrated that hypothalamic stem cells can control the rate of whole organism aging through the release of exosomes into the cerebrospinal fluid [175]. This shows that EVs from a small cell population can achieve systemic-level effects related to aging. Aging is marked by molecular and cellular changes that underlie the aging process, leading to tissue dysfunction and increased mortality risk. EVs from various cell types are now recognized as modulators of these aging-related processes. Here, we highlight recent findings regarding the role of EVs in aging.

Cellular senescence is a process that halts cell proliferation in response to damage, contributing to aging-related conditions [176]. Senescent cells release detrimental factors to the extracellular environment through the SASP [177]. EVs play a crucial role in these processes, contributing to EV-associated SASP (evSASP) [177]. EVs from healthy, non-senescent cells can decrease senescence by modifying the extracellular environment, thus reducing inflammation and improving tissue function [178–180].

Oxidative stress, particularly the control of cellular redox status, is implicated in aging and age-related diseases [5, 181]. EVs have been shown to regulate redox status by transferring molecules with antioxidant properties, protecting cells from oxidative damage [182].

As we have stated, chronic inflammation is a key process during aging and contributes to age-related diseases [17]. EVs can modulate the immune system, with some EVs having anti-inflammatory and immunomodulatory properties [183], decreasing pro-inflammatory factors and potentially reducing the effects of inflammaging [174, 184].

In summary, EVs may serve as a therapeutic tool to address the molecular and cellular changes associated with aging, potentially enhancing tissue regeneration and modulating aging-related processes (Fig. 8.4).

Challenges and Ethical Considerations

While the potential of targeting cell communication for aging interventions is great, there are several significant challenges and ethical considerations that need to be addressed before reaching humans.

One of the foremost challenges is ensuring the safety of these interventions. As the field of aging interventions expands, many people are questioning the use of therapies only tested on preclinical models. The use of young plasma exchange, for example, has raised ethical concerns. Similarly, the use of EVs as therapeutics requires rigorous testing to determine their long-term effects and to minimize unintended consequences, such as off-target effects and immune responses. The development and clinical use of these interventions will require regulatory approval, and as aging itself is not considered a disease, the regulatory framework should change, or otherwise try to apply the findings to specific age-related diseases. Access to these innovative therapies may be limited by cost, location, or other factors, raising questions about equitable access to potential anti-aging treatments.

In conclusion, as the field of aging interventions advances, it is essential to address these challenges and ethical considerations to ensure that these therapies are developed and applied responsibly.

Senotherapy: Senolytics and Senomorphics

The process of aging is accompanied by alterations in cellular communication, which can lead to the development of age-related diseases and a decline in tissue and organ function. Senolytics and senomorphics are two distinct yet interconnected approaches aimed at modulating these age-related changes in cell communication (Fig. 8.4).

Senolytics

Senolytics are compounds or drugs designed to selectively target and eliminate senescent cells. With age, senescent cells accumulate in the organism because they are resistant to apoptosis and the immune system fails to eliminate them. Recent studies have shown that the enhanced survival of senescent cells depends on the activation of various mechanisms referred to as senescent cell anti-apoptotic pathways (SCAPs). SCAPs are directly related to many features of the senescent phenotype such as the Senescence-Associated Growth Arrest (SAGA), the SASP, the unfolded protein response (UPR), and the PI3K-Akt signaling [185, 186]. Elimination of senescent cells has the potential to delay aging, treat age-related diseases, and extend healthspan. Importantly, SCAPs are now under investigation as targets for senolytic drugs. Indeed, senolytics work by inducing apoptosis (programmed cell death) in senescent cells, thereby reducing their detrimental effects on surrounding tissues [187].

Approaches used to identify senolytics include bioinformatic analysis of SCAPs for drug development and screening of drug libraries on different senescent cell types in culture [188]. Several key SCAPs identified include tyrosine kinases/ephrins, BCL-2/BCL-xL antiapoptotic family members, PI3K/AKT, p53/p21/PAI-1&2, and HIF-1α [187]. The first senolytic agents were those drugs that could target several SCAPs at the same time, such as desatinib, quercetin, curcumin, and fisetin [189]. These natural compounds have safe profiles and can be administered orally. To date, the senolytic effect of desatinib + quercetin (D + Q) has been tested in many in vivo models where intermittent treatment alleviated several age-related diseases such as progeroid symptoms, physical dysfunction, osteoporosis, hepatic steatosis, insulin resistance, neurodegeneration, vasomotor dysfunction, pulmonary fibrosis, chronic kidney disease, and skeletal muscle dysfunction [190]. Based on the beneficial effects observed in vitro and in vivo, the combination of D + Q has progressed to human clinical trials for idiopathic pulmonary fibrosis (NCT02874989), chronic kidney disease (NCT02848131), skeletal health (NCT04313634), hematopoietic stem cell transplant survivors (NCT02652052), Alzheimer's disease (NCT04063124) and cancer survivors (NCT04733534). However, due to the heterogeneity of SCAPs, not all senescent cells were sensitive to these compounds. Since then, many senolytic drugs have been developed, including those that target a single SCAP, such as Navitoclax (ABT-263) which induces apoptosis in a restricted range of senescent cell types [191, 192]. Navitoclax is an inhibitor that targets BCL-2 and BCL-xL by binding to the BH3 binding groove, thereby disrupting their interactions with the pro-death protein BIM and initiating apoptosis [193]. However, Navitoclax is not able to induce apoptosis in all senescent cells and has severe side effects, including thrombocytopenia and neutropenia [194].

Second-generation senolytics are now being identified using high-throughput analysis [195, 196]. Among them, cardiac glycosides [197], galactose-modified prodrugs [198], HSP90 inhibitors [199], molecules targeting the p53 pathway such as FoxO4-p53 inhibitors [200], MDM2/p53 inhibitors [201] and ubiquitin-specific peptidase 7 (USP7) inhibitors [202]. Finally, an alternative approach for designing safer senolytics is based on the innovative proteolysis-targeting chimera (PROTAC) technology. PROTACs are comprised of three elements: a ligand of a target protein, a linker, and an E3 ligase recruiting ligand. Given the many targetable proteins or protein–protein interactions involved in the regulation of senescence, the increasing number of E3 ligase binding moieties, and the linker diversity, it is probable that many PROTAC senolytics will be developed in the future [203–205].

Taken together, senolytics function through various mechanisms, including inhibiting pro-survival pathways that protect senescent cells from apoptosis and activating apoptotic pathways specifically in senescent cells. Importantly, senolytics have shown promise in preclinical studies and animal models for various age-related conditions, such as osteoarthritis [206], atherosclerosis [207], and neurodegenerative diseases [208, 209]. Clinical trials are underway to evaluate their safety and efficacy in humans [210–212].

Senomorphics

Senomorphics are compounds or interventions that do not directly eliminate senescent cells but instead modify their secretory profile, shifting it from pro-inflammatory to a more youthful, pro-regenerative state. Senomorphics suppress the detrimental effects of SASP components secreted by senescent cells without causing cell death. Thus, senomorphics aim to rejuvenate senescent cells and promote a healthier cellular environment. The first senomorphics were mainly discovered by serendipity [213], including rapamycin [214, 215], metformin [216–218], resveratrol [219, 220] and aspirin [221, 222].

In general, senomorphics modulate senescence by suppressing SASP production and release via targeting NF-κB, mTOR, IL-1α, p38MAPK, and other signaling pathways [223–226]. Therefore, new senomorphics include NF-κB inhibitors [227, 228], p38MAPK inhibitors [229, 230], JAK/STAT inhibitors [231, 232], and ATM inhibitors [233, 234]. A second approach is based on the neutralization of specific SASP factors such as IL-1α, IL-8, and IL-6 with specific antibodies [235]. Indeed, a monoclonal antibody against IL-8 (ABX-IL-8) is known to impair IL-8 signaling [236] and is used as a SASP modulator. Likewise, another monoclonal antibody against IL-6 (Mab-IL-6.8) blocks IL-6 and reduces its effects [237].

In this manner, senomorphics may be a preferred method of intervention as senescent cells are also implicated in physiological processes such as wound healing and development [31, 238]. Complete blockage of SASP can also be detrimental since several SASP components participate in various intercellular signaling pathways and therefore, SASP modulation may also negatively affect those functions. Finally, senomorphics do not permanently eliminate the source of SASP, and therefore chronic/repetitive treatment is needed, which may increase the incidence of toxic side effects. Consequently, caution should be attributed regarding the timeline of senomorphics administration.

Future Directions

Unanswered Questions

Despite significant progress in understanding the role of intercellular communication in aging, numerous unanswered questions remain, highlighting opportunities for future research. For example, there is a need to elucidate how the dysregulation of intercellular communication uniquely impacts different tissues and organ systems during aging, and which tissues are more sensitive to the effect of changes in the extracellular environment. There is also a lack of knowledge on the dynamics of these intercellular signaling processes across the lifespan, it will be important to identify transition points during aging and potential intervention windows. Furthermore, an emerging area of interest involves exploring the microbiome's influence on the alteration of the extracellular environment during aging and determining how gut microbiota and their metabolites shape signaling pathways. What is clear is that we need to better understand how cell-intrinsic changes that occur during aging affect the extracellular environment and cell–cell signaling. Although we are starting to know which extracellular factors change with age, we still lack the basic molecular mechanisms that ultimately lead to the dysregulation of intercellular communication during aging.

Technological Advances

To address the unanswered questions and advance our understanding of cellular communication in aging, researchers are increasingly taking advantage of cutting-edge technologies. For instance, single-cell omics, which include transcriptomics, proteomics, and epigenomics, have revolutionized our ability to dissect individual cells within complex tissues, providing insights into unique cell subpopulations and their role in aging-related processes. More specifically, researchers have characterized the transcriptomic changes that occur with heterochronic parabiosis at single-cell resolution [156, 157], leading to a better understanding of changes that occur in specific cell types during parabiosis. Microfluidic "organ-on-a-chip" models reproduce the structure and function of human organs, enabling the study of intercellular communication in a physiologically relevant context and bridging the gap between in vitro and in vivo studies [239]. The drastic increase in omics research has led to the creation of vast datasets, making it imperative to develop adequate machine learning and computational biology approaches. The integration of data from genomics, transcriptomics, proteomics, and metabolomics through multi-omics approaches will doubtless provide a more comprehensive understanding of both aging and how cells interact with each other within this process.

Conclusion

Aging is not a simple linear process but rather a multifaceted phenomenon influenced by a myriad of mechanisms including those related to cell-to-cell communication mechanisms. These include direct communication (juxtacrine) and indirect communication (autocrine, paracrine, and endocrine). Understanding the complexity of these interactions is essential to comprehend the aging process. Dysregulated communication can contribute to various age-related diseases, including neurodegenerative disorders, cardiovascular diseases, and cancer. Deciphering the signaling pathways involved can provide crucial insights into disease prevention and treatment. By modulating these communication networks, researchers and clinicians can develop innovative strategies to mitigate the effects of aging and age-related diseases.

In conclusion, the chapter on cell-to-cell communication in aging elucidates the essential role of intercellular signaling in the aging process. This knowledge opens up exciting possibilities for therapeutic interventions that could potentially extend human lifespan and improve the quality of life in old age. The pursuit of these prospects promises to be a dynamic and evolving field of research with the potential to revolutionize our approach to aging and age-related diseases.

References

1. Viña J, Borrás C, Miquel J. Theories of ageing. IUBMB Life. 2007;59(4–5):249-54. doi: https://doi.org/10.1080/15216540601178067.
2. Guo J, Huang X, Dou L, Yan M, Shen T, Tang W, et al. Aging and aging-related diseases: from molecular mechanisms to interventions and treatments. Signal Transduct Target Ther. 2022;7(1):391. doi: https://doi.org/10.1038/s41392-022-01251-0.
3. Partridge L. The new biology of ageing. Philosophical transactions of the Royal Society of London Series B, Biological sciences. 2010;365(1537):147–54. doi: https://doi.org/10.1098/rstb.2009.0222.
4. Kirkwood TB. Understanding the odd science of aging. Cell. 2005;120(4):437–47. doi: https://doi.org/10.1016/j.cell.2005.01.027.
5. Lopez-Otin C, Blasco MA, Partridge L, Serrano M, Kroemer G. Hallmarks of aging: An expanding universe. Cell. 2023;186(2):243–78. doi: https://doi.org/10.1016/j.cell.2022.11.001.
6. Song D, Yang D, Powell CA, Wang X. Cell–cell communication: old mystery and new opportunity. Cell Biology and Toxicology. 2019;35(2):89–93. doi: https://doi.org/10.1007/S10565-019-09470-Y.
7. Mittelbrunn M, Sánchez-Madrid F. 'Intercellular communication: diverse structures for exchange of genetic information'. Nature reviews Molecular cell biology. 2012;13(5):328-. doi: https://doi.org/10.1038/NRM3335.
8. Armingol E, Officer A, Harismendy O, Lewis NE. Deciphering cell–cell interactions and communication from gene expression. Nature Reviews Genetics. 2021;22(2):1-. doi: https://doi.org/10.1038/S41576-020-00292-X.
9. Alberts B, Hunt T, Johnson A, Lewis J, Morgan D, Raff MC, et al. Molecular biology of the cell. Sixth edition. 6th . ed. New York, NY: Garland Science, Taylor and Francis Group; 2015.
10. Iizuka K, Machida T, Hirafuji M. Skeletal muscle is an endocrine organ. J Pharmacol Sci. 2014;125(2):125–31. doi: https://doi.org/10.1254/jphs.14r02cp.
11. Zouboulis CC. The skin as an endocrine organ. Dermatoendocrinol. 2009;1(5):250–2. doi: https://doi.org/10.4161/derm.1.5.9499.
12. Yang BA, Westerhof TM, Sabin K, Merajver SD, Aguilar CA. Engineered Tools to Study Intercellular Communication. Advanced Science. 2021;8(3). doi: https://doi.org/10.1002/ADVS.202002825.
13. Van Helvert S, Storm C, Friedl P. Mechanoreciprocity in cell migration. Nature cell biology. 2018;20(1):8-. doi: https://doi.org/10.1038/S41556-017-0012-0.
14. Swartz MA, Tschumperlin DJ, Kamm RD, Drazen JM. Mechanical stress is communicated between different cell types to elicit matrix remodeling. Proceedings of the National Academy of Sciences of the United States of America. 2001;98(11):6180-. doi: https://doi.org/10.1073/PNAS.111133298.
15. Pakshir P, Alizadehgiashi M, Wong B, Coelho NM, Chen X, Gong Z, et al. Dynamic fibroblast contractions attract remote macrophages in fibrillar collagen matrix. Nature communications. 2019;10(1). doi: https://doi.org/10.1038/S41467-019-09709-6.
16. Fafián-Labora JA, O'Loghlen A. Classical and Nonclassical Intercellular Communication in Senescence and Ageing. Trends in cell biology. 2020;30(8):628–39. doi: https://doi.org/10.1016/J.TCB.2020.05.003.
17. Franceschi C, Campisi J. Chronic inflammation (inflammaging) and its potential contribution to age-associated diseases. The journals of gerontology Series A, Biological sciences and medical sciences. 2014;69 Suppl 1:S4–9. doi: https://doi.org/10.1093/gerona/glu057.
18. Fafian-Labora JA, O'Loghlen A. NF-kappaB/IKK activation by small extracellular vesicles within the SASP. Aging cell. 2021;20(7):e13426. doi: https://doi.org/10.1111/acel.13426.
19. Mittelbrunn M, Sanchez-Madrid F. Intercellular communication: diverse structures for exchange of genetic information. Nature reviews Molecular cell biology. 2012;13(5):328–35. doi: https://doi.org/10.1038/nrm3335.

20. Fontana L, Partridge L, Longo VD. Extending healthy life span--from yeast to humans. Science. 2010;328(5976):321–6. doi: https://doi.org/10.1126/science.1172539.
21. Russell SJ, Kahn CR. Endocrine regulation of ageing. Nature reviews Molecular cell biology. 2007;8(9):681–91. doi: https://doi.org/10.1038/nrm2234.
22. Franceschi C, Bonafè M, Valensin S, Olivieri F, De Luca M, Ottaviani E, et al. Inflamm-aging. An evolutionary perspective on immunosenescence. Annals of the New York Academy of Sciences. 2000;908:244–54. doi: https://doi.org/10.1111/j.1749-6632.2000.tb06651.x.
23. Lefebvre JS, Maue AC, Eaton SM, Lanthier PA, Tighe M, Haynes L. The aged microenvironment contributes to the age-related functional defects of CD4 T cells in mice. Aging cell. 2012;11(5):732–40. doi: https://doi.org/10.1111/j.1474-9726.2012.00836.x.
24. Sarnak MJ, Levey AS, Schoolwerth AC, Coresh J, Culleton B, Hamm LL, et al. Kidney disease as a risk factor for development of cardiovascular disease: a statement from the American Heart Association Councils on Kidney in Cardiovascular Disease, High Blood Pressure Research, Clinical Cardiology, and Epidemiology and Prevention. Hypertension (Dallas, Tex : 1979). 2003;42(5):1050–65. doi: https://doi.org/10.1161/01.HYP.0000102971.85504.7c.
25. El Mai M, Bird M, Allouche A, Targen S, Serifoglu N, Lopes-Bastos B, et al. Gut-specific telomerase expression counteracts systemic aging in telomerase-deficient zebrafish. Nat Aging. 2023;3(5):567–84. doi: https://doi.org/10.1038/s43587-023-00401-5.
26. Durieux J, Wolff S, Dillin A. The cell-non-autonomous nature of electron transport chain-mediated longevity. Cell. 2011;144(1):79–91. doi: https://doi.org/10.1016/j.cell.2010.12.016.
27. Hernandez-Segura A, Nehme J, Demaria M. Hallmarks of Cellular Senescence. Trends in cell biology. 2018;28(6):436–53. doi: https://doi.org/10.1016/j.tcb.2018.02.001.
28. Antonangeli F, Zingoni A, Soriani A, Santoni A. Senescent cells: Living or dying is a matter of NK cells. J Leukoc Biol. 2019;105(6):1275–83. doi: https://doi.org/10.1002/jlb.mr0718-299r.
29. Kang TW, Yevsa T, Woller N, Hoenicke L, Wuestefeld T, Dauch D, et al. Senescence surveillance of pre-malignant hepatocytes limits liver cancer development. Nature. 2011;479(7374):547–51. doi: https://doi.org/10.1038/nature10599.
30. Andrade AM, Sun M, Gasek NS, Hargis GR, Sharafieh R, Xu M. Role of Senescent Cells in Cutaneous Wound Healing. Biology (Basel). 2022;11(12). doi: https://doi.org/10.3390/biology11121731.
31. Demaria M, Ohtani N, Youssef SA, Rodier F, Toussaint W, Mitchell JR, et al. An essential role for senescent cells in optimal wound healing through secretion of PDGF-AA. Developmental cell. 2014;31(6):722–33. doi: https://doi.org/10.1016/j.devcel.2014.11.012.
32. Freund A, Orjalo AV, Desprez PY, Campisi J. Inflammatory networks during cellular senescence: causes and consequences. Trends Mol Med. 2010;16(5):238–46. doi: https://doi.org/10.1016/j.molmed.2010.03.003.
33. He S, Sharpless NE. Senescence in Health and Disease. Cell. 2017;169(6):1000–11. doi: https://doi.org/10.1016/j.cell.2017.05.015.
34. Franco AC, Aveleira C, Cavadas C. Skin senescence: mechanisms and impact on whole-body aging. Trends Mol Med. 2022;28(2):97–109. doi: https://doi.org/10.1016/j.molmed.2021.12.003.
35. Victorelli S, Lagnado A, Halim J, Moore W, Talbot D, Barrett K, et al. Senescent human melanocytes drive skin ageing via paracrine telomere dysfunction. The EMBO journal. 2019;38(23):e101982. doi: https://doi.org/10.15252/embj.2019101982.
36. Musi N, Valentine JM, Sickora KR, Baeuerle E, Thompson CS, Shen Q, et al. Tau protein aggregation is associated with cellular senescence in the brain. Aging cell. 2018;17(6):e12840. doi: https://doi.org/10.1111/acel.12840.
37. Bhat R, Crowe EP, Bitto A, Moh M, Katsetos CD, Garcia FU, et al. Astrocyte senescence as a component of Alzheimer's disease. PLoS One. 2012;7(9):e45069. doi: https://doi.org/10.1371/journal.pone.0045069.
38. Russo T, Riessland M. Age-Related Midbrain Inflammation and Senescence in Parkinson's Disease. Front Aging Neurosci. 2022;14:917797. doi: https://doi.org/10.3389/fnagi.2022.917797.

39. Trindade M, Oigman W, Fritsch Neves M. Potential Role of Endothelin in Early Vascular Aging. Curr Hypertens Rev. 2017;13(1):33–40. doi: https://doi.org/10.2174/1573402113666170414165735.
40. Schmidt-Ott KM, Kagiyama S, Phillips MI. The multiple actions of angiotensin II in atherosclerosis. Regul Pept. 2000;93(1–3):65–77. doi: https://doi.org/10.1016/s0167-0115(00)00178-6.
41. Csiszar A, Sosnowska D, Wang M, Lakatta EG, Sonntag WE, Ungvari Z. Age-associated proinflammatory secretory phenotype in vascular smooth muscle cells from the non-human primate Macaca mulatta: reversal by resveratrol treatment. The journals of gerontology Series A, Biological sciences and medical sciences. 2012;67(8):811–20. doi: https://doi.org/10.1093/gerona/glr228.
42. Buendia P, Montes de Oca A, Madueno JA, Merino A, Martin-Malo A, Aljama P, et al. Endothelial microparticles mediate inflammation-induced vascular calcification. FASEB journal : official publication of the Federation of American Societies for Experimental Biology. 2015;29(1):173–81. doi: https://doi.org/10.1096/fj.14-249706.
43. Zhang L, Xu Z, Wu Y, Liao J, Zeng F, Shi L. Akt/eNOS and MAPK signaling pathways mediated the phenotypic switching of thoracic aorta vascular smooth muscle cells in aging/hypertensive rats. Physiol Res. 2018;67(4):543-53. doi: https://doi.org/10.33549/physiolres.933779.
44. Nakano-Kurimoto R, Ikeda K, Uraoka M, Nakagawa Y, Yutaka K, Koide M, et al. Replicative senescence of vascular smooth muscle cells enhances the calcification through initiating the osteoblastic transition. American journal of physiology Heart and circulatory physiology. 2009;297(5):H1673-84. doi: https://doi.org/10.1152/ajpheart.00455.2009.
45. Lombardi M, Mantione ME, Baccellieri D, Ferrara D, Castellano R, Chiesa R, et al. P2X7 receptor antagonism modulates IL-1beta and MMP9 in human atherosclerotic vessels. Sci Rep. 2017;7(1):4872. doi: https://doi.org/10.1038/s41598-017-05137-y.
46. Wu CM, Zheng L, Wang Q, Hu YW. The emerging role of cell senescence in atherosclerosis. Clin Chem Lab Med. 2020;59(1):27–38. doi: https://doi.org/10.1515/cclm-2020-0601.
47. He Y, Xie W, Li H, Jin H, Zhang Y, Li Y. Cellular Senescence in Sarcopenia: Possible Mechanisms and Therapeutic Potential. Front Cell Dev Biol. 2021;9:793088. doi: https://doi.org/10.3389/fcell.2021.793088.
48. da Silva PFL, Ogrodnik M, Kucheryavenko O, Glibert J, Miwa S, Cameron K, et al. The bystander effect contributes to the accumulation of senescent cells in vivo. Aging cell. 2019;18(1):e12848. doi: https://doi.org/10.1111/acel.12848.
49. Wan M, Gray-Gaillard EF, Elisseeff JH. Cellular senescence in musculoskeletal homeostasis, diseases, and regeneration. Bone Res. 2021;9(1):41. doi: https://doi.org/10.1038/s41413-021-00164-y.
50. Farr JN, Fraser DG, Wang H, Jaehn K, Ogrodnik MB, Weivoda MM, et al. Identification of Senescent Cells in the Bone Microenvironment. J Bone Miner Res. 2016;31(11):1920-9. doi: https://doi.org/10.1002/jbmr.2892.
51. Piemontese M, Almeida M, Robling AG, Kim HN, Xiong J, Thostenson JD, et al. Old age causes de novo intracortical bone remodeling and porosity in mice. JCI insight. 2017;2(17). doi: https://doi.org/10.1172/jci.insight.93771.
52. Chen Q, Liu K, Robinson AR, Clauson CL, Blair HC, Robbins PD, et al. DNA damage drives accelerated bone aging via an NF-kappaB-dependent mechanism. J Bone Miner Res. 2013;28(5):1214–28. doi: https://doi.org/10.1002/jbmr.1851.
53. Kim HN, Xiong J, MacLeod RS, Iyer S, Fujiwara Y, Cawley KM, et al. Osteocyte RANKL is required for cortical bone loss with age and is induced by senescence. JCI insight. 2020;5(19). doi: https://doi.org/10.1172/jci.insight.138815.
54. Thomas R, Wang W, Su DM. Contributions of Age-Related Thymic Involution to Immunosenescence and Inflammaging. Immun Ageing. 2020;17:2. doi: https://doi.org/10.1186/s12979-020-0173-8.
55. Franceschi C, Santoro A, Capri M. The complex relationship between Immunosenescence and Inflammaging: Special issue on the New Biomedical Perspectives. Semin Immunopathol. vol 52020. p. 517–20.

56. Pawelec G, Adibzadeh M, Pohla H, Schaudt K. Immunosenescence: ageing of the immune system. Immunol Today. vol 9. England1995. p. 420–2.
57. Pawelec G. Age and immunity: What is “immunosenescence”? Exp Gerontol. 2018;105:4–9. doi: https://doi.org/10.1016/j.exger.2017.10.024.
58. Olivieri F, Prattichizzo F, Grillari J, Balistreri CR. Cellular Senescence and Inflammaging in Age-Related Diseases. Mediators of Inflammation. 2018;2018. doi: https://doi.org/10.1155/2018/9076485.
59. Nikolich-Zugich J. The twilight of immunity: emerging concepts in aging of the immune system. Nat Immunol. 2018;19(1):10–9. doi: https://doi.org/10.1038/s41590-017-0006-x.
60. Fulop T, Witkowski JM, Pawelec G, Alan C, Larbi A. On the immunological theory of aging. Interdiscip Top Gerontol. 2014;39:163–76. doi: https://doi.org/10.1159/000358904.
61. Flores KG, Li J, Sempowski GD, Haynes BF, Hale LP. Analysis of the human thymic perivascular space during aging. The Journal of clinical investigation. 1999;104(8):1031–9. doi: https://doi.org/10.1172/JCI7558.
62. Lynch HE, Goldberg GL, Chidgey A, Van den Brink MR, Boyd R, Sempowski GD. Thymic involution and immune reconstitution. Trends Immunol. 2009;30(7):366–73. doi: https://doi.org/10.1016/j.it.2009.04.003.
63. Tamir A, Eisenbraun MD, Garcia GG, Miller RA. Age-dependent alterations in the assembly of signal transduction complexes at the site of T cell/APC interaction. Journal of immunology (Baltimore, Md : 1950). 2000;165(3):1243–51. doi: https://doi.org/10.4049/jimmunol.165.3.1243.
64. Linton PJ, Haynes L, Klinman NR, Swain SL. Antigen-independent changes in naive CD4 T cells with aging. The Journal of experimental medicine. 1996;184(5):1891–900. doi: https://doi.org/10.1084/jem.184.5.1891.
65. Haynes L, Linton PJ, Eaton SM, Tonkonogy SL, Swain SL. Interleukin 2, but not other common gamma chain-binding cytokines, can reverse the defect in generation of CD4 effector T cells from naive T cells of aged mice. The Journal of experimental medicine. 1999;190(7):1013–24. doi: https://doi.org/10.1084/jem.190.7.1013.
66. Eaton SM, Burns EM, Kusser K, Randall TD, Haynes L. Age-related defects in CD4 T cell cognate helper function lead to reductions in humoral responses. The Journal of experimental medicine. 2004;200(12):1613–22. doi: https://doi.org/10.1084/jem.20041395.
67. Haynes L, Eaton SM. The effect of age on the cognate function of CD4+ T cells. Immunol Rev. 2005;205:220–8. doi: https://doi.org/10.1111/j.0105-2896.2005.00255.x.
68. Goronzy JJ, Weyand CM. T cell development and receptor diversity during aging. Curr Opin Immunol. 2005;17(5):468–75. doi: https://doi.org/10.1016/j.coi.2005.07.020.
69. Murasko DM, Jiang J. Response of aged mice to primary virus infections. Immunol Rev. 2005;205:285–96. doi: https://doi.org/10.1111/j.0105-2896.2005.00273.x.
70. Messaoudi I, Lemaoult J, Guevara-Patino JA, Metzner BM, Nikolich-Zugich J. Age-related CD8 T cell clonal expansions constrict CD8 T cell repertoire and have the potential to impair immune defense. The Journal of experimental medicine. 2004;200(10):1347–58. doi: https://doi.org/10.1084/jem.20040437.
71. Kapasi ZF, Murali-Krishna K, McRae ML, Ahmed R. Defective generation but normal maintenance of memory T cells in old mice. Eur J Immunol. 2002;32(6):1567–73. doi: https://doi.org/10.1002/1521-4141(200206)32:6<1567::AID-IMMU1567>3.0.CO;2-P.
72. Solana R, Tarazona R, Gayoso I, Lesur O, Dupuis G, Fulop T. Innate immunosenescence: effect of aging on cells and receptors of the innate immune system in humans. Semin Immunol. 2012;24(5):331–41. doi: https://doi.org/10.1016/j.smim.2012.04.008.
73. Fortin CF, McDonald PP, Lesur O, Fulop T, Jr. Aging and neutrophils: there is still much to do. Rejuvenation Res. 2008;11(5):873–82. doi: https://doi.org/10.1089/rej.2008.0750.
74. Nyugen J, Agrawal S, Gollapudi S, Gupta S. Impaired functions of peripheral blood monocyte subpopulations in aged humans. J Clin Immunol. 2010;30(6):806–13. doi: https://doi.org/10.1007/s10875-010-9448-8.
75. Agrawal A, Agrawal S, Cao JN, Su H, Osann K, Gupta S. Altered innate immune functioning of dendritic cells in elderly humans: a role of phosphoinositide 3-kinase-signaling pathway.

Journal of immunology (Baltimore, Md : 1950). 2007;178(11):6912–22. doi: https://doi.org/10.4049/jimmunol.178.11.6912.
76. Ray D, Yung R. Immune senescence, epigenetics and autoimmunity. Clin Immunol. 2018;196:59–63. doi: https://doi.org/10.1016/j.clim.2018.04.002.
77. Biagi E, Nylund L, Candela M, Ostan R, Bucci L, Pini E, et al. Through ageing, and beyond: gut microbiota and inflammatory status in seniors and centenarians. PLoS One. 2010;5(5):e10667. doi: https://doi.org/10.1371/journal.pone.0010667.
78. Sanada F, Taniyama Y, Muratsu J, Otsu R, Shimizu H, Rakugi H, et al. Source of Chronic Inflammation in Aging. Front Cardiovasc Med. 2018;5:12. doi: https://doi.org/10.3389/fcvm.2018.00012.
79. Mangiola F, Nicoletti A, Gasbarrini A, Ponziani FR. Gut microbiota and aging. European review for medical and pharmacological sciences. 2018;22(21):7404–13. doi: https://doi.org/10.26355/eurrev_201811_16280.
80. Roh JS, Sohn DH. Damage-Associated Molecular Patterns in Inflammatory Diseases. Immune Netw. 2018;18(4):e27. doi: https://doi.org/10.4110/in.2018.18.e27.
81. Sochocka M, Diniz BS, Leszek J. Inflammatory Response in the CNS: Friend or Foe? Mol Neurobiol. 2017;54(10):8071–89. doi: https://doi.org/10.1007/s12035-016-0297-1.
82. Fagiolo U, Cossarizza A, Scala E, Fanales-Belasio E, Ortolani C, Cozzi E, et al. Increased cytokine production in mononuclear cells of healthy elderly people. Eur J Immunol. 1993;23(9):2375–8. doi: https://doi.org/10.1002/eji.1830230950.
83. Ferrucci L, Fabbri E. Inflammageing: chronic inflammation in ageing, cardiovascular disease, and frailty. Nature reviews Cardiology. 2018;15(9):505–22. doi: https://doi.org/10.1038/s41569-018-0064-2.
84. Sayed N, Huang Y, Nguyen K, Krejciova-Rajaniemi Z, Grawe AP, Gao T, et al. An inflammatory aging clock (iAge) based on deep learning tracks multimorbidity, immunosenescence, frailty and cardiovascular aging. Nat Aging. 2021;1:598–615. doi: https://doi.org/10.1038/s43587-021-00082-y.
85. Zhang B, Bailey WM, Braun KJ, Gensel JC. Age decreases macrophage IL-10 expression: Implications for functional recovery and tissue repair in spinal cord injury. Exp Neurol. 2015;273:83–91. doi: https://doi.org/10.1016/j.expneurol.2015.08.001.
86. Bulati M, Buffa S, Candore G, Caruso C, Dunn-Walters DK, Pellicano M, et al. B cells and immunosenescence: a focus on IgG+IgD-CD27- (DN) B cells in aged humans. Ageing Res Rev. 2011;10(2):274–84. doi: https://doi.org/10.1016/j.arr.2010.12.002.
87. Claesson MJ, Cusack S, O'Sullivan O, Greene-Diniz R, de Weerd H, Flannery E, et al. Composition, variability, and temporal stability of the intestinal microbiota of the elderly. Proceedings of the National Academy of Sciences of the United States of America. 2011;108 Suppl 1(Suppl 1):4586–91. doi: https://doi.org/10.1073/pnas.1000097107.
88. Ljungquist B, Berg S, Lanke J, McClearn GE, Pedersen NL. The effect of genetic factors for longevity: a comparison of identical and fraternal twins in the Swedish Twin Registry. The journals of gerontology Series A, Biological sciences and medical sciences. 1998;53(6):M441–6. doi: https://doi.org/10.1093/gerona/53a.6.m441.
89. Soerensen M. Genetic variation and human longevity. Dan Med J. 2012;59(5):B4454.
90. Pilling LC, Atkins JL, Bowman K, Jones SE, Tyrrell J, Beaumont RN, et al. Human longevity is influenced by many genetic variants: evidence from 75,000 UK Biobank participants. Aging. 2016;8(3):547–60. doi: https://doi.org/10.18632/aging.100930.
91. Hasler R, Venkatesh G, Tan Q, Flachsbart F, Sinha A, Rosenstiel P, et al. Genetic interplay between human longevity and metabolic pathways - a large-scale eQTL study. Aging cell. 2017;16(4):716–25. doi: https://doi.org/10.1111/acel.12598.
92. Maynard S, Fang EF, Scheibye-Knudsen M, Croteau DL, Bohr VA. DNA Damage, DNA Repair, Aging, and Neurodegeneration. Cold Spring Harb Perspect Med. 2015;5(10). doi: https://doi.org/10.1101/cshperspect.a025130.
93. Lombard DB, Chua KF, Mostoslavsky R, Franco S, Gostissa M, Alt FW. DNA repair, genome stability, and aging. Cell. 2005;120(4):497–512. doi: https://doi.org/10.1016/j.cell.2005.01.028.

94. McKinnon PJ. DNA repair deficiency and neurological disease. Nat Rev Neurosci. 2009;10(2):100–12. doi: https://doi.org/10.1038/nrn2559.
95. Jeppesen DK, Bohr VA, Stevnsner T. DNA repair deficiency in neurodegeneration. Progress in neurobiology. 2011;94(2):166–200. doi: https://doi.org/10.1016/j.pneurobio.2011.04.013.
96. Debrabant B, Soerensen M, Flachsbart F, Dato S, Mengel-From J, Stevnsner T, et al. Human longevity and variation in DNA damage response and repair: study of the contribution of sub-processes using competitive gene-set analysis. Eur J Hum Genet. 2014;22(9):1131–6. doi: https://doi.org/10.1038/ejhg.2013.299.
97. Lee JM, Yang PW, Yang SY, Chuang TH, Tung EC, Chen JS, et al. Genetic variants in DNA repair predicts the survival of patients with esophageal cancer. Ann Surg. 2011;253(5):918–27. doi: https://doi.org/10.1097/SLA.0b013e318216f374.
98. Wang S, Wang J, Bai Y, Wang Q, Liu L, Zhang K, et al. The genetic variations in DNA repair genes ERCC2 and XRCC1 were associated with the overall survival of advanced non-small-cell lung cancer patients. Cancer medicine. 2016;5(9):2332–42. doi: https://doi.org/10.1002/cam4.822.
99. Aubert G, Lansdorp PM. Telomeres and aging. Physiological reviews. 2008;88(2):557–79. doi: https://doi.org/10.1152/physrev.00026.2007.
100. Victorelli S, Passos JF. Telomeres and Cell Senescence - Size Matters Not. EBioMedicine. 2017;21:14–20. doi: https://doi.org/10.1016/j.ebiom.2017.03.027.
101. Baird DM, Rowson J, Wynford-Thomas D, Kipling D. Extensive allelic variation and ultra-short telomeres in senescent human cells. Nat Genet. 2003;33(2):203–7. doi: https://doi.org/10.1038/ng1084.
102. Liston A, Humblet-Baron S, Duffy D, Goris A. Human immune diversity: from evolution to modernity. Nat Immunol. 2021;22(12):1479–89. doi: https://doi.org/10.1038/s41590-021-01058-1.
103. Mangino M, Roederer M, Beddall MH, Nestle FO, Spector TD. Innate and adaptive immune traits are differentially affected by genetic and environmental factors. Nature communications. 2017;8:13850. doi: https://doi.org/10.1038/ncomms13850.
104. Brodin P, Jojic V, Gao T, Bhattacharya S, Angel CJ, Furman D, et al. Variation in the human immune system is largely driven by non-heritable influences. Cell. 2015;160(1–2):37–47. doi: https://doi.org/10.1016/j.cell.2014.12.020.
105. Li Y, Oosting M, Smeekens SP, Jaeger M, Aguirre-Gamboa R, Le KTT, et al. A Functional Genomics Approach to Understand Variation in Cytokine Production in Humans. Cell. 2016;167(4):1099–110 e14. doi: https://doi.org/10.1016/j.cell.2016.10.017.
106. Pairo-Castineira E, Clohisey S, Klaric L, Bretherick AD, Rawlik K, Pasko D, et al. Genetic mechanisms of critical illness in COVID-19. Nature. 2021;591(7848):92–8. doi: https://doi.org/10.1038/s41586-020-03065-y.
107. Initiative C-HG. Mapping the human genetic architecture of COVID-19. Nature. 2021;600(7889):472–7. doi: https://doi.org/10.1038/s41586-021-03767-x.
108. King DE, Mainous AG, 3rd, Matheson EM, Everett CJ. Impact of healthy lifestyle on mortality in people with normal blood pressure, LDL cholesterol, and C-reactive protein. Eur J Prev Cardiol. 2013;20(1):73–9. doi: https://doi.org/10.1177/1741826711425776.
109. Leitao C, Mignano A, Estrela M, Fardilha M, Figueiras A, Roque F, et al. The Effect of Nutrition on Aging-A Systematic Review Focusing on Aging-Related Biomarkers. Nutrients. 2022;14(3). doi: https://doi.org/10.3390/nu14030554.
110. Martini D. Health Benefits of Mediterranean Diet. Nutrients. 2019;11(8). doi: https://doi.org/10.3390/nu11081802.
111. Tyrovolas S, Haro JM, Mariolis A, Piscopo S, Valacchi G, Tsakountakis N, et al. Successful aging, dietary habits and health status of elderly individuals: a k-dimensional approach within the multi-national MEDIS study. Exp Gerontol. 2014;60:57–63. doi: https://doi.org/10.1016/j.exger.2014.09.010.
112. Galbete C, Schwingshackl L, Schwedhelm C, Boeing H, Schulze MB. Evaluating Mediterranean diet and risk of chronic disease in cohort studies: an umbrella review of meta-analyses. Eur J Epidemiol. 2018;33(10):909–31. doi: https://doi.org/10.1007/s10654-018-0427-3.

113. Johnson SC. Nutrient Sensing, Signaling and Ageing: The Role of IGF-1 and mTOR in Ageing and Age-Related Disease. Subcell Biochem. 2018;90:49–97. doi: https://doi.org/10.1007/978-981-13-2835-0_3.
114. Longo VD, Antebi A, Bartke A, Barzilai N, Brown-Borg HM, Caruso C, et al. Interventions to Slow Aging in Humans: Are We Ready? Aging cell. 2015;14(4):497–510. doi: https://doi.org/10.1111/acel.12338.
115. Mannick JB, Lamming DW. Targeting the biology of aging with mTOR inhibitors. Nat Aging. 2023;3(6):642–60. doi: https://doi.org/10.1038/s43587-023-00416-y.
116. Papadopoli D, Boulay K, Kazak L, Pollak M, Mallette F, Topisirovic I, et al. mTOR as a central regulator of lifespan and aging. F1000Res. 2019;8. doi: https://doi.org/10.12688/f1000research.17196.1.
117. Bonkowski MS, Sinclair DA. Slowing ageing by design: the rise of NAD(+) and sirtuin-activating compounds. Nature reviews Molecular cell biology. 2016;17(11):679–90. doi: https://doi.org/10.1038/nrm.2016.93.
118. Grabowska W, Sikora E, Bielak-Zmijewska A. Sirtuins, a promising target in slowing down the ageing process. Biogerontology. 2017;18(4):447–76. doi: https://doi.org/10.1007/s10522-017-9685-9.
119. Fabrizio P, Pozza F, Pletcher SD, Gendron CM, Longo VD. Regulation of longevity and stress resistance by Sch9 in yeast. Science. 2001;292(5515):288–90. doi: https://doi.org/10.1126/science.1059497.
120. Jia K, Chen D, Riddle DL. The TOR pathway interacts with the insulin signaling pathway to regulate C. elegans larval development, metabolism and life span. Development. 2004;131(16):3897–906. doi: https://doi.org/10.1242/dev.01255.
121. Bjedov I, Toivonen JM, Kerr F, Slack C, Jacobson J, Foley A, et al. Mechanisms of life span extension by rapamycin in the fruit fly Drosophila melanogaster. Cell metabolism. 2010;11(1):35–46. doi: https://doi.org/10.1016/j.cmet.2009.11.010.
122. Harrison DE, Strong R, Sharp ZD, Nelson JF, Astle CM, Flurkey K, et al. Rapamycin fed late in life extends lifespan in genetically heterogeneous mice. Nature. 2009;460(7253):392–5. doi: https://doi.org/10.1038/nature08221.
123. Dai G, Li Y, Zhang M, Lu P, Zhang Y, Wang H, et al. The Regulation of the AMPK/mTOR Axis Mitigates Tendon Stem/Progenitor Cell Senescence and Delays Tendon Aging. Stem Cell Rev Rep. 2023;19(5):1492–506. doi: https://doi.org/10.1007/s12015-023-10526-0.
124. Saoudaoui S, Bernard M, Cardin GB, Malaquin N, Christopoulos A, Rodier F. mTOR as a senescence manipulation target: A forked road. Advances in cancer research. 2021;150:335–63. doi: https://doi.org/10.1016/bs.acr.2021.02.002.
125. Kennedy BK, Austriaco NR, Jr., Zhang J, Guarente L. Mutation in the silencing gene SIR4 can delay aging in S. cerevisiae. Cell. 1995;80(3):485–96. doi: https://doi.org/10.1016/0092-8674(95)90499-9.
126. Tissenbaum HA, Guarente L. Increased dosage of a sir-2 gene extends lifespan in Caenorhabditis elegans. Nature. 2001;410(6825):227–30. doi: https://doi.org/10.1038/35065638.
127. Rogina B, Helfand SL. Sir2 mediates longevity in the fly through a pathway related to calorie restriction. Proceedings of the National Academy of Sciences of the United States of America. 2004;101(45):15998–6003. doi: https://doi.org/10.1073/pnas.0404184101.
128. Kanfi Y, Naiman S, Amir G, Peshti V, Zinman G, Nahum L, et al. The sirtuin SIRT6 regulates lifespan in male mice. Nature. 2012;483(7388):218–21. doi: https://doi.org/10.1038/nature10815.
129. Lee SH, Lee JH, Lee HY, Min KJ. Sirtuin signaling in cellular senescence and aging. BMB Rep. 2019;52(1):24–34. doi: https://doi.org/10.5483/BMBRep.2019.52.1.290.
130. Mas-Bargues C, Vina-Almunia J, Ingles M, Sanz-Ros J, Gambini J, Ibanez-Cabellos JS, et al. Role of p16INK4a and BMI-1 in oxidative stress-induced premature senescence in human dental pulp stem cells. Redox Biology. 2017;12:690–8. doi: https://doi.org/10.1016/j.redox.2017.04.002.
131. Liao N, Shi Y, Zhang C, Zheng Y, Wang Y, Zhao B, et al. Antioxidants inhibit cell senescence and preserve stemness of adipose tissue-derived stem cells by reducing ROS generation during

long-term in vitro expansion. Stem cell research & therapy. 2019;10(1):306. doi: https://doi.org/10.1186/s13287-019-1404-9.
132. Jeong JH, Kim MB, Kim C, Hwang JK. Inhibitory effect of vitamin C on intrinsic aging in human dermal fibroblasts and hairless mice. Food Sci Biotechnol. 2018;27(2):555–64. doi: https://doi.org/10.1007/s10068-017-0252-6.
133. La Fata G, Seifert N, Weber P, Mohajeri MH. Vitamin E Supplementation Delays Cellular Senescence In Vitro. Biomed Res Int. 2015;2015:563247. doi: https://doi.org/10.1155/2015/563247.
134. Flanagan EW, Most J, Mey JT, Redman LM. Calorie Restriction and Aging in Humans. Annu Rev Nutr. 2020;40:105–33. doi: https://doi.org/10.1146/annurev-nutr-122319-034601.
135. Giacomello E, Toniolo L. The Potential of Calorie Restriction and Calorie Restriction Mimetics in Delaying Aging: Focus on Experimental Models. Nutrients. 2021;13(7). doi: https://doi.org/10.3390/nu13072346.
136. Finley LW, Haigis MC. The coordination of nuclear and mitochondrial communication during aging and calorie restriction. Ageing Res Rev. 2009;8(3):173–88. doi: https://doi.org/10.1016/j.arr.2009.03.003.
137. Gabande-Rodriguez E, Gomez de Las Heras MM, Mittelbrunn M. Control of Inflammation by Calorie Restriction Mimetics: On the Crossroad of Autophagy and Mitochondria. Cells. 2019;9(1). doi: https://doi.org/10.3390/cells9010082.
138. Mattison JA, Colman RJ, Beasley TM, Allison DB, Kemnitz JW, Roth GS, et al. Caloric restriction improves health and survival of rhesus monkeys. Nature communications. 2017;8:14063. doi: https://doi.org/10.1038/ncomms14063.
139. Rachakatla A, Kalashikam RR. Calorie Restriction-Regulated Molecular Pathways and Its Impact on Various Age Groups: An Overview. DNA Cell Biol. 2022;41(5):459–68. doi: https://doi.org/10.1089/dna.2021.0922.
140. Redman LM, Smith SR, Burton JH, Martin CK, Il'yasova D, Ravussin E. Metabolic Slowing and Reduced Oxidative Damage with Sustained Caloric Restriction Support the Rate of Living and Oxidative Damage Theories of Aging. Cell metabolism. 2018;27(4):805–15 e4. doi: https://doi.org/10.1016/j.cmet.2018.02.019.
141. Chung KW, Chung HY. The Effects of Calorie Restriction on Autophagy: Role on Aging Intervention. Nutrients. 2019;11(12). doi: https://doi.org/10.3390/nu11122923.
142. Luevano-Martinez LA, Forni MF, Peloggia J, Watanabe IS, Kowaltowski AJ. Calorie restriction promotes cardiolipin biosynthesis and distribution between mitochondrial membranes. Mechanisms of ageing and development. 2017;162:9–17. doi: https://doi.org/10.1016/j.mad.2017.02.004.
143. Derous D, Mitchell SE, Wang L, Green CL, Wang Y, Chen L, et al. The effects of graded levels of calorie restriction: XI. Evaluation of the main hypotheses underpinning the life extension effects of CR using the hepatic transcriptome. Aging. 2017;9(7):1770–824. doi: https://doi.org/10.18632/aging.101269.
144. Ning YC, Cai GY, Zhuo L, Gao JJ, Dong D, Cui S, et al. Short-term calorie restriction protects against renal senescence of aged rats by increasing autophagic activity and reducing oxidative damage. Mechanisms of ageing and development. 2013;134(11–12):570–9. doi: https://doi.org/10.1016/j.mad.2013.11.006.
145. Singh H, Kaur T, Manchanda S, Kaur G. Intermittent fasting combined with supplementation with Ayurvedic herbs reduces anxiety in middle aged female rats by anti-inflammatory pathways. Biogerontology. 2017;18(4):601–14. doi: https://doi.org/10.1007/s10522-017-9706-8.
146. Vasconcelos AR, Yshii LM, Viel TA, Buck HS, Mattson MP, Scavone C, et al. Intermittent fasting attenuates lipopolysaccharide-induced neuroinflammation and memory impairment. J Neuroinflammation. 2014;11:85. doi: https://doi.org/10.1186/1742-2094-11-85.
147. Jung KJ, Lee EK, Kim JY, Zou Y, Sung B, Heo HS, et al. Effect of short term calorie restriction on pro-inflammatory NF-kB and AP-1 in aged rat kidney. Inflamm Res. 2009;58(3):143–50. doi: https://doi.org/10.1007/s00011-008-7227-2.

148. Meydani SN, Das SK, Pieper CF, Lewis MR, Klein S, Dixit VD, et al. Long-term moderate calorie restriction inhibits inflammation without impairing cell-mediated immunity: a randomized controlled trial in non-obese humans. Aging. 2016;8(7):1416–31. doi: https://doi.org/10.18632/aging.100994.
149. Fontana L, Nehme J, Demaria M. Caloric restriction and cellular senescence. Mechanisms of ageing and development. 2018;176:19–23. doi: https://doi.org/10.1016/j.mad.2018.10.005.
150. Kurki E, Shi J, Martonen E, Finckenberg P, Mervaala E. Distinct effects of calorie restriction on adipose tissue cytokine and angiogenesis profiles in obese and lean mice. Nutr Metab (Lond). 2012;9(1):64. doi: https://doi.org/10.1186/1743-7075-9-64.
151. Park CY, Park S, Kim MS, Kim HK, Han SN. Effects of mild calorie restriction on lipid metabolism and inflammation in liver and adipose tissue. Biochemical and biophysical research communications. 2017;490(3):636–42. doi: https://doi.org/10.1016/j.bbrc.2017.06.090.
152. McCay CM, Pope F, Lunsford W, Sperling G, Sambhavaphol P. Parabiosis between old and young rats. Gerontologia. 1957;1(1):7–17. doi: https://doi.org/10.1159/000210677.
153. Conboy IM, Rando TA. Heterochronic parabiosis for the study of the effects of aging on stem cells and their niches. Cell cycle (Georgetown, Tex). 2012;11(12):2260–7. doi: https://doi.org/10.4161/cc.20437.
154. Conboy IM, Conboy MJ, Wagers AJ, Girma ER, Weissman IL, Rando TA. Rejuvenation of aged progenitor cells by exposure to a young systemic environment. Nature. 2005;433(7027):760–4. doi: https://doi.org/10.1038/nature03260.
155. Loffredo FS, Steinhauser ML, Jay SM, Gannon J, Pancoast JR, Yalamanchi P, et al. Growth differentiation factor 11 is a circulating factor that reverses age-related cardiac hypertrophy. Cell. 2013;153(4):828–39. doi: https://doi.org/10.1016/j.cell.2013.04.015.
156. Ma S, Wang S, Ye Y, Ren J, Chen R, Li W, et al. Heterochronic parabiosis induces stem cell revitalization and systemic rejuvenation across aged tissues. Cell Stem Cell. 2022;29(6):990–1005 e10. doi: https://doi.org/10.1016/j.stem.2022.04.017.
157. Palovics R, Keller A, Schaum N, Tan W, Fehlmann T, Borja M, et al. Molecular hallmarks of heterochronic parabiosis at single-cell resolution. Nature. 2022;603(7900):309–14. doi: https://doi.org/10.1038/s41586-022-04461-2.
158. Ballak DB, Brunt VE, Sapinsley ZJ, Ziemba BP, Richey JJ, Zigler MC, et al. Short-term interleukin-37 treatment improves vascular endothelial function, endurance exercise capacity, and whole-body glucose metabolism in old mice. Aging cell. 2020;19(1):e13074. doi: https://doi.org/10.1111/acel.13074.
159. Frohlich J, Vinciguerra M. Candidate rejuvenating factor GDF11 and tissue fibrosis: friend or foe? Geroscience. 2020;42(6):1475–98. doi: https://doi.org/10.1007/s11357-020-00279-w.
160. Mera P, Laue K, Ferron M, Confavreux C, Wei J, Galan-Diez M, et al. Osteocalcin Signaling in Myofibers Is Necessary and Sufficient for Optimum Adaptation to Exercise. Cell metabolism. 2016;23(6):1078–92. doi: https://doi.org/10.1016/j.cmet.2016.05.004.
161. Thery C, Ostrowski M, Segura E. Membrane vesicles as conveyors of immune responses. Nature reviews Immunology. 2009;9(8):581–93. doi: https://doi.org/10.1038/nri2567.
162. Zoller M. Tetraspanins: push and pull in suppressing and promoting metastasis. Nature reviews Cancer. 2009;9(1):40–55. doi: https://doi.org/10.1038/nrc2543.
163. Raposo G, Stoorvogel W. Extracellular vesicles: exosomes, microvesicles, and friends. The Journal of cell biology. 2013;200(4):373–83. doi: https://doi.org/10.1083/jcb.201211138.
164. Lee RH, Pulin AA, Seo MJ, Kota DJ, Ylostalo J, Larson BL, et al. Intravenous hMSCs improve myocardial infarction in mice because cells embolized in lung are activated to secrete the anti-inflammatory protein TSG-6. Cell Stem Cell. 2009;5(1):54–63. doi: https://doi.org/10.1016/j.stem.2009.05.003.
165. Eggenhofer E, Benseler V, Kroemer A, Popp FC, Geissler EK, Schlitt HJ, et al. Mesenchymal stem cells are short-lived and do not migrate beyond the lungs after intravenous infusion. Front Immunol. 2012;3:297. doi: https://doi.org/10.3389/fimmu.2012.00297.
166. Toma C, Wagner WR, Bowry S, Schwartz A, Villanueva F. Fate of culture-expanded mesenchymal stem cells in the microvasculature: in vivo observations of cell kinetics.

Circulation research. 2009;104(3):398–402. doi: https://doi.org/10.1161/CIRCRESAHA.108.187724.
167. Chen L, Tredget EE, Wu PY, Wu Y. Paracrine factors of mesenchymal stem cells recruit macrophages and endothelial lineage cells and enhance wound healing. PLoS One. 2008;3(4):e1886. doi: https://doi.org/10.1371/journal.pone.0001886.
168. De Jong OG, Van Balkom BW, Schiffelers RM, Bouten CV, Verhaar MC. Extracellular vesicles: potential roles in regenerative medicine. Front Immunol. 2014;5:608. doi: https://doi.org/10.3389/fimmu.2014.00608.
169. Grigorian-Shamagian L, Liu W, Fereydooni S, Middleton RC, Valle J, Cho JH, et al. Cardiac and systemic rejuvenation after cardiosphere-derived cell therapy in senescent rats. European heart journal. 2017;38(39):2957–67. doi: https://doi.org/10.1093/eurheartj/ehx454.
170. Kulkarni R, Bajaj M, Ghode S, Jalnapurkar S, Limaye L, Kale VP. Intercellular Transfer of Microvesicles from Young Mesenchymal Stromal Cells Rejuvenates Aged Murine Hematopoietic Stem Cells. Stem cells (Dayton, Ohio). 2018;36(3):420–33. doi: https://doi.org/10.1002/stem.2756.
171. Oh M, Lee J, Kim YJ, Rhee WJ, Park JH. Exosomes Derived from Human Induced Pluripotent Stem Cells Ameliorate the Aging of Skin Fibroblasts. International journal of molecular sciences. 2018;19(6). doi: https://doi.org/10.3390/ijms19061715.
172. Liu S, Mahairaki V, Bai H, Ding Z, Li J, Witwer KW, et al. Highly Purified Human Extracellular Vesicles Produced by Stem Cells Alleviate Aging Cellular Phenotypes of Senescent Human Cells. Stem cells (Dayton, Ohio). 2019;37(6):779–90. doi: https://doi.org/10.1002/stem.2996.
173. Lee BR, Kim JH, Choi ES, Cho JH, Kim E. Effect of young exosomes injected in aged mice. Int J Nanomedicine. 2018;13:5335–45. doi: https://doi.org/10.2147/IJN.S170680.
174. Wang W, Wang L, Ruan L, Oh J, Dong X, Zhuge Q, et al. Extracellular vesicles extracted from young donor serum attenuate inflammaging via partially rejuvenating aged T-cell immunotolerance. FASEB journal : official publication of the Federation of American Societies for Experimental Biology. 2018;32(11):fj201800059R. doi: https://doi.org/10.1096/fj.201800059R.
175. Zhang Y, Kim MS, Jia B, Yan J, Zuniga-Hertz JP, Han C, et al. Hypothalamic stem cells control ageing speed partly through exosomal miRNAs. Nature. 2017;548(7665):52–7. doi: https://doi.org/10.1038/nature23282.
176. Campisi J, d'Adda di Fagagna F. Cellular senescence: when bad things happen to good cells. Nature Reviews Molecular Cell Biology. 2007;8(9):729–40. doi: https://doi.org/10.1038/nrm2233.
177. Coppé JP, Desprez PY, Krtolica A, Campisi J. The senescence-associated secretory phenotype: the dark side of tumor suppression. Annu Rev Pathol. 2010;5:99–118. doi: https://doi.org/10.1146/annurev-pathol-121808-102144.
178. Mas-Bargues C, Sanz-Ros J, Roman-Dominguez A, Gimeno-Mallench L, Ingles M, Vina J, et al. Extracellular Vesicles from Healthy Cells Improves Cell Function and Stemness in Premature Senescent Stem Cells by miR-302b and HIF-1alpha Activation. Biomolecules. 2020;10(6). doi: https://doi.org/10.3390/biom10060957.
179. Fafián-Labora JA, Rodríguez-Navarro JA, O'Loghlen A. Small Extracellular Vesicles Have GST Activity and Ameliorate Senescence-Related Tissue Damage. Cell metabolism. 2020;32(1):71–86.e5. doi: https://doi.org/10.1016/j.cmet.2020.06.004.
180. Sanz-Ros J, Romero-Garcia N, Mas-Bargues C, Monleon D, Gordevicius J, Brooke RT, et al. Small extracellular vesicles from young adipose-derived stem cells prevent frailty, improve health span, and decrease epigenetic age in old mice. Sci Adv. 2022;8(42):eabq2226. doi: https://doi.org/10.1126/sciadv.abq2226.
181. Vina J, Borras C, Abdelaziz KM, Garcia-Valles R, Gomez-Cabrera MC. The free radical theory of aging revisited: the cell signaling disruption theory of aging. Antioxidants & redox signaling. 2013;19(8):779–87. doi: https://doi.org/10.1089/ars.2012.5111.
182. Mas-Bargues C, Sanz-Ros J, Romero-Garcia N, Huete-Acevedo J, Dromant M, Borras C. Small extracellular vesicles from senescent stem cells trigger adaptive mechanisms in young stem cells by increasing antioxidant enzyme expression. Redox Biol. 2023;62:102668. doi: https://doi.org/10.1016/j.redox.2023.102668.

183. Bruno S, Deregibus MC, Camussi G. The secretome of mesenchymal stromal cells: Role of extracellular vesicles in immunomodulation. Immunol Lett. 2015;168(2):154–8. doi: https://doi.org/10.1016/j.imlet.2015.06.007.
184. Dorronsoro A, Santiago FE, Grassi D, Zhang T, Lai RC, McGowan SJ, et al. Mesenchymal stem cell-derived extracellular vesicles reduce senescence and extend health span in mouse models of aging. Aging cell. 2021;20(4):e13337. doi: https://doi.org/10.1111/acel.13337.
185. Soto-Gamez A, Quax WJ, Demaria M. Regulation of Survival Networks in Senescent Cells: From Mechanisms to Interventions. J Mol Biol. 2019;431(15):2629–43. doi: https://doi.org/10.1016/j.jmb.2019.05.036.
186. Childs BG, Baker DJ, Kirkland JL, Campisi J, van Deursen JM. Senescence and apoptosis: dueling or complementary cell fates? EMBO Reports. 2014;15(11):1139–53. doi: https://doi.org/10.15252/embr.201439245.
187. Zhu Y, Tchkonia T, Pirtskhalava T, Gower AC, Ding H, Giorgadze N, et al. The Achilles' heel of senescent cells: from transcriptome to senolytic drugs. Aging cell. 2015;14(4):644–58. doi: https://doi.org/10.1111/acel.12344.
188. Zhang L, Pitcher LE, Prahalad V, Niedernhofer LJ, Robbins PD. Recent advances in the discovery of senolytics. Mechanisms of ageing and development. 2021;200:111587. doi: https://doi.org/10.1016/j.mad.2021.111587.
189. Zhu Y, Doornebal EJ, Pirtskhalava T, Giorgadze N, Wentworth M, Fuhrmann-Stroissnigg H, et al. New agents that target senescent cells: the flavone, fisetin, and the BCL-XL inhibitors, A1331852 and A1155463. Aging. 2017;9(3):955–63. doi: https://doi.org/10.18632/aging.101202.
190. Wissler Gerdes EO, Zhu Y, Tchkonia T, Kirkland JL. Discovery, development, and future application of senolytics: theories and predictions. The FEBS journal. 2020;287(12):2418–27. doi: https://doi.org/10.1111/febs.15264.
191. Zhu Y, Tchkonia T, Fuhrmann-Stroissnigg H, Dai HM, Ling YY, Stout MB, et al. Identification of a novel senolytic agent, navitoclax, targeting the Bcl-2 family of anti-apoptotic factors. Aging cell. 2016;15(3):428–35. doi: https://doi.org/10.1111/acel.12445.
192. Chang J, Wang Y, Shao L, Laberge RM, Demaria M, Campisi J, et al. Clearance of senescent cells by ABT263 rejuvenates aged hematopoietic stem cells in mice. Nature medicine. 2016;22(1):78–83. doi: https://doi.org/10.1038/nm.4010.
193. Tse C, Shoemaker AR, Adickes J, Anderson MG, Chen J, Jin S, et al. ABT-263: a potent and orally bioavailable Bcl-2 family inhibitor. Cancer research. 2008;68(9):3421–8. doi: https://doi.org/10.1158/0008-5472.CAN-07-5836.
194. Schoenwaelder SM, Jarman KE, Gardiner EE, Hua M, Qiao J, White MJ, et al. Bcl-xL-inhibitory BH3 mimetics can induce a transient thrombocytopathy that undermines the hemostatic function of platelets. Blood. 2011;118(6):1663–74. doi: https://doi.org/10.1182/blood-2011-04-347849.
195. Fuhrmann-Stroissnigg H, Santiago FE, Grassi D, Ling Y, Niedernhofer LJ, Robbins PD. SA-beta-Galactosidase-Based Screening Assay for the Identification of Senotherapeutic Drugs. Journal of visualized experiments : JoVE. 2019(148). doi: https://doi.org/10.3791/58133.
196. Robbins PD, Jurk D, Khosla S, Kirkland JL, LeBrasseur NK, Miller JD, et al. Senolytic Drugs: Reducing Senescent Cell Viability to Extend Health Span. Annu Rev Pharmacol Toxicol. 2021;61:779–803. doi: https://doi.org/10.1146/annurev-pharmtox-050120-105018.
197. Triana-Martinez F, Picallos-Rabina P, Da Silva-Alvarez S, Pietrocola F, Llanos S, Rodilla V, et al. Identification and characterization of Cardiac Glycosides as senolytic compounds. Nature communications. 2019;10(1):4731. doi: https://doi.org/10.1038/s41467-019-12888-x.
198. Guerrero A, Guiho R, Herranz N, Uren A, Withers DJ, Martinez-Barbera JP, et al. Galactose-modified duocarmycin prodrugs as senolytics. Aging cell. 2020;19(4):e13133. doi: https://doi.org/10.1111/acel.13133.
199. Fuhrmann-Stroissnigg H, Ling YY, Zhao J, McGowan SJ, Zhu Y, Brooks RW, et al. Identification of HSP90 inhibitors as a novel class of senolytics. Nature communications. 2017;8(1):422. doi: https://doi.org/10.1038/s41467-017-00314-z.

200. Zhang C, Xie Y, Chen H, Lv L, Yao J, Zhang M, et al. FOXO4-DRI alleviates age-related testosterone secretion insufficiency by targeting senescent Leydig cells in aged mice. Aging. 2020;12(2):1272–84. doi: https://doi.org/10.18632/aging.102682.
201. Chae JB, Jang H, Son C, Park CW, Choi H, Jin S, et al. Targeting senescent retinal pigment epithelial cells facilitates retinal regeneration in mouse models of age-related macular degeneration. Geroscience. 2021;43(6):2809–33. doi: https://doi.org/10.1007/s11357-021-00457-4.
202. He Y, Li W, Lv D, Zhang X, Zhang X, Ortiz YT, et al. Inhibition of USP7 activity selectively eliminates senescent cells in part via restoration of p53 activity. Aging cell. 2020;19(3):e13117. doi: https://doi.org/10.1111/acel.13117.
203. Wakita M, Takahashi A, Sano O, Loo TM, Imai Y, Narukawa M, et al. A BET family protein degrader provokes senolysis by targeting NHEJ and autophagy in senescent cells. Nature communications. 2020;11(1):1935. doi: https://doi.org/10.1038/s41467-020-15719-6.
204. Burslem GM, Crews CM. Proteolysis-Targeting Chimeras as Therapeutics and Tools for Biological Discovery. Cell. 2020;181(1):102–14. doi: https://doi.org/10.1016/j.cell.2019.11.031.
205. Jia Y, Han L, Ramage CL, Wang Z, Weng CC, Yang L, et al. Co-targeting BCL-XL and BCL-2 by PROTAC 753B eliminates leukemia cells and enhances efficacy of chemotherapy by targeting senescent cells. Haematologica. 2023;108(10):2626–38. doi: https://doi.org/10.3324/haematol.2022.281915.
206. Gil TH, Zheng H, Lee HG, Shin JW, Hwang SW, Jang KM, et al. Senolytic drugs relieve pain by reducing peripheral nociceptive signaling without modifying joint tissue damage in spontaneous osteoarthritis. Aging. 2022;14(15):6006–27. doi: https://doi.org/10.18632/aging.204204.
207. Garrido AM, Kaistha A, Uryga AK, Oc S, Foote K, Shah A, et al. Efficacy and limitations of senolysis in atherosclerosis. Cardiovascular research. 2022;118(7):1713–27. doi: https://doi.org/10.1093/cvr/cvab208.
208. Ogrodnik M, Evans SA, Fielder E, Victorelli S, Kruger P, Salmonowicz H, et al. Whole-body senescent cell clearance alleviates age-related brain inflammation and cognitive impairment in mice. Aging cell. 2021;20(2):e13296. doi: https://doi.org/10.1111/acel.13296.
209. Lee S, Wang EY, Steinberg AB, Walton CC, Chinta SJ, Andersen JK. A guide to senolytic intervention in neurodegenerative disease. Mechanisms of ageing and development. 2021;200:111585. doi: https://doi.org/10.1016/j.mad.2021.111585.
210. Gonzales MM, Garbarino VR, Marques Zilli E, Petersen RC, Kirkland JL, Tchkonia T, et al. Senolytic Therapy to Modulate the Progression of Alzheimer's Disease (SToMP-AD): A Pilot Clinical Trial. J Prev Alzheimers Dis. 2022;9(1):22–9. doi: https://doi.org/10.14283/jpad.2021.62.
211. Nambiar A, Kellogg D, 3rd, Justice J, Goros M, Gelfond J, Pascual R, et al. Senolytics dasatinib and quercetin in idiopathic pulmonary fibrosis: results of a phase I, single-blind, single-center, randomized, placebo-controlled pilot trial on feasibility and tolerability. EBioMedicine. 2023;90:104481. doi: https://doi.org/10.1016/j.ebiom.2023.104481.
212. Justice JN, Nambiar AM, Tchkonia T, LeBrasseur NK, Pascual R, Hashmi SK, et al. Senolytics in idiopathic pulmonary fibrosis: Results from a first-in-human, open-label, pilot study. EBioMedicine. 2019;40:554–63. doi: https://doi.org/10.1016/j.ebiom.2018.12.052.
213. Zhang L, Pitcher LE, Prahalad V, Niedernhofer LJ, Robbins PD. Targeting cellular senescence with senotherapeutics: senolytics and senomorphics. The FEBS journal. 2023;290(5):1362–83. doi: https://doi.org/10.1111/febs.16350.
214. Selvarani R, Mohammed S, Richardson A. Effect of rapamycin on aging and age-related diseases-past and future. Geroscience. 2021;43(3):1135–58. doi: https://doi.org/10.1007/s11357-020-00274-1.
215. Lamming DW, Ye L, Sabatini DM, Baur JA. Rapalogs and mTOR inhibitors as anti-aging therapeutics. The Journal of clinical investigation. 2013;123(3):980–9. doi: https://doi.org/10.1172/JCI64099.

216. Fang J, Yang J, Wu X, Zhang G, Li T, Wang X, et al. Metformin alleviates human cellular aging by upregulating the endoplasmic reticulum glutathione peroxidase 7. Aging cell. 2018;17(4):e12765. doi: https://doi.org/10.1111/acel.12765.
217. Zhang C, Chen M, Zhou N, Qi Y. Metformin Prevents H(2)O(2)-Induced Senescence in Human Lens Epithelial B3 Cells. Med Sci Monit Basic Res. 2020;26:e923391. doi: https://doi.org/10.12659/MSMBR.923391.
218. Smieszek A, Strek Z, Kornicka K, Grzesiak J, Weiss C, Marycz K. Antioxidant and Anti-Senescence Effect of Metformin on Mouse Olfactory Ensheathing Cells (mOECs) May Be Associated with Increased Brain-Derived Neurotrophic Factor Levels-An Ex Vivo Study. International journal of molecular sciences. 2017;18(4). doi: https://doi.org/10.3390/ijms18040872.
219. Xia L, Wang XX, Hu XS, Guo XG, Shang YP, Chen HJ, et al. Resveratrol reduces endothelial progenitor cells senescence through augmentation of telomerase activity by Akt-dependent mechanisms. Br J Pharmacol. 2008;155(3):387–94. doi: https://doi.org/10.1038/bjp.2008.272.
220. Ali D, Chen L, Kowal JM, Okla M, Manikandan M, AlShehri M, et al. Resveratrol inhibits adipocyte differentiation and cellular senescence of human bone marrow stromal stem cells. Bone. 2020;133:115252. doi: https://doi.org/10.1016/j.bone.2020.115252.
221. Bode-Boger SM, Martens-Lobenhoffer J, Tager M, Schroder H, Scalera F. Aspirin reduces endothelial cell senescence. Biochemical and biophysical research communications. 2005;334(4):1226–32. doi: https://doi.org/10.1016/j.bbrc.2005.07.014.
222. Feng M, Kim J, Field K, Reid C, Chatzistamou I, Shim M. Aspirin ameliorates the long-term adverse effects of doxorubicin through suppression of cellular senescence. FASEB Bioadv. 2019;1(9):579–90. doi: https://doi.org/10.1096/fba.2019-00041.
223. Fryer LG, Parbu-Patel A, Carling D. The Anti-diabetic drugs rosiglitazone and metformin stimulate AMP-activated protein kinase through distinct signaling pathways. J Biol Chem. 2002;277(28):25226–32. doi: https://doi.org/10.1074/jbc.M202489200.
224. Moiseeva O, Deschenes-Simard X, St-Germain E, Igelmann S, Huot G, Cadar AE, et al. Metformin inhibits the senescence-associated secretory phenotype by interfering with IKK/NF-kappaB activation. Aging cell. 2013;12(3):489–98. doi: https://doi.org/10.1111/acel.12075.
225. Pitozzi V, Mocali A, Laurenzana A, Giannoni E, Cifola I, Battaglia C, et al. Chronic resveratrol treatment ameliorates cell adhesion and mitigates the inflammatory phenotype in senescent human fibroblasts. The journals of gerontology Series A, Biological sciences and medical sciences. 2013;68(4):371–81. doi: https://doi.org/10.1093/gerona/gls183.
226. Alimbetov D, Davis T, Brook AJ, Cox LS, Faragher RG, Nurgozhin T, et al. Suppression of the senescence-associated secretory phenotype (SASP) in human fibroblasts using small molecule inhibitors of p38 MAP kinase and MK2. Biogerontology. 2016;17(2):305–15. doi: https://doi.org/10.1007/s10522-015-9610-z.
227. Tilstra JS, Robinson AR, Wang J, Gregg SQ, Clauson CL, Reay DP, et al. NF-kappaB inhibition delays DNA damage-induced senescence and aging in mice. The Journal of clinical investigation. 2012;122(7):2601–12. doi: https://doi.org/10.1172/JCI45785.
228. Zhang L, Zhao J, Mu X, McGowan SJ, Angelini L, O'Kelly RD, et al. Novel small molecule inhibition of IKK/NF-kappaB activation reduces markers of senescence and improves healthspan in mouse models of aging. Aging cell. 2021;20(12):e13486. doi: https://doi.org/10.1111/acel.13486.
229. Freund A, Patil CK, Campisi J. p38MAPK is a novel DNA damage response-independent regulator of the senescence-associated secretory phenotype. The EMBO journal. 2011;30(8):1536–48. doi: https://doi.org/10.1038/emboj.2011.69.
230. Wang Y, Liu L, Zhou D. Inhibition of p38 MAPK attenuates ionizing radiation-induced hematopoietic cell senescence and residual bone marrow injury. Radiat Res. 2011;176(6):743–52. doi: https://doi.org/10.1667/rr2727.1.
231. Griveau A, Wiel C, Ziegler DV, Bergo MO, Bernard D. The JAK1/2 inhibitor ruxolitinib delays premature aging phenotypes. Aging cell. 2020;19(4):e13122. doi: https://doi.org/10.1111/acel.13122.

232. Wu W, Fu J, Gu Y, Wei Y, Ma P, Wu J. JAK2/STAT3 regulates estrogen-related senescence of bone marrow stem cells. The Journal of endocrinology. 2020;245(1):141–53. doi: https://doi.org/10.1530/JOE-19-0518.
233. Zhao J, Zhang L, Lu A, Han Y, Colangelo D, Bukata C, et al. ATM is a key driver of NF-kappaB-dependent DNA-damage-induced senescence, stem cell dysfunction and aging. Aging. 2020;12(6):4688–710. doi: https://doi.org/10.18632/aging.102863.
234. Kuk MU, Kim JW, Lee YS, Cho KA, Park JT, Park SC. Alleviation of Senescence via ATM Inhibition in Accelerated Aging Models. Mol Cells. 2019;42(3):210–7. doi: https://doi.org/10.14348/molcells.2018.0352.
235. Lagoumtzi SM, Chondrogianni N. Senolytics and senomorphics: Natural and synthetic therapeutics in the treatment of aging and chronic diseases. Free radical biology & medicine. 2021;171:169–90. doi: https://doi.org/10.1016/j.freeradbiomed.2021.05.003.
236. Yang XD, Corvalan JR, Wang P, Roy CM, Davis CG. Fully human anti-interleukin-8 monoclonal antibodies: potential therapeutics for the treatment of inflammatory disease states. J Leukoc Biol. 1999;66(3):401–10. doi: https://doi.org/10.1002/jlb.66.3.401.
237. Kuilman T, Michaloglou C, Vredeveld LC, Douma S, van Doorn R, Desmet CJ, et al. Oncogene-induced senescence relayed by an interleukin-dependent inflammatory network. Cell. 2008;133(6):1019–31. doi: https://doi.org/10.1016/j.cell.2008.03.039.
238. Muñoz-Espín D, Cañamero M, Maraver A, Gómez-López G, Contreras J, Murillo-Cuesta S, et al. Programmed cell senescence during mammalian embryonic development. Cell. 2013;155(5):1104–18. doi: https://doi.org/10.1016/j.cell.2013.10.019.
239. Ingber DE. Human organs-on-chips for disease modelling, drug development and personalized medicine. Nat Rev Genet. 2022;23(8):467–91. doi: https://doi.org/10.1038/s41576-022-00466-9.

Chapter 9
The Senescence-Associated Secretory Phenotype: Induction, Regulation, Function and Therapeutic Interventions to Counteract the Negative Effects

Daniela Frasca

Aging and Inflammaging

Aging is a complex phenomenon due to stochastic, genetic, epigenetic and environmental causes [1]. Aging is characterized by chronic low-grade systemic inflammation known as "inflammaging" [2], which is a risk factor for morbidity and mortality of elderly individuals because it is implicated in the development and pathogenicity of several diseases of old age, such as cancer [3, 4], type-2 diabetes mellitus (T2DM) [5], heart disease [6], Alzheimer's disease [7], rheumatoid arthritis [8], Parkinson's [9], multiple sclerosis [10], atherosclerosis [11], age-related macular degeneration [12], and osteoporosis [13]. Serum markers of inflammaging are inflammatory mediators (C-reactive proteins, IL-6, TNF-α, IFN-γ, mannose-binding lectin) [1, 2, 14], as well as metabolic markers (triglycerides, cholesterol, low- and high-density lipoproteins, serum amyloid A, glucose, albumin, alanine aminotransferase, creatinine, non-esterified fatty acids) [15].

Inflammaging is due to several causes. Examples are polymorphisms in the promoter regions of pro- [16–19] and anti-inflammatory genes [20]; chronic stimulation of immune system cells with viruses [21, 22]; stimuli from the gut microbiota [23] or from mitochondrial DNA released in the extracellular space and functioning as a damage-associated molecular pattern [24, 25]; increased intestinal permeability [26]; increased fat mass with age [27–30]; circulating micro-RNAs [31]. Inflammaging is also caused by the accumulation of senescent cells in the body of aged

D. Frasca (✉)
Department of Microbiology and Immunology, University of Miami Miller School of Medicine, Miami, FL, USA
e-mail: dfrasca@med.miami.edu

Sylvester Comprehensive Cancer Center, University of Miami Miller School of Medicine, Miami, FL, USA

V. Bueno (ed.), *Cellular and Molecular Aspects of Ageing*,
https://doi.org/10.1007/978-3-031-55022-5_9

individuals, mainly due to impaired immune surveillance and impaired cytotoxicity by NK cells [32]. Senescent cells promote tissue degeneration, malignant transformation [33, 34], and lead to the development of inflammatory conditions and diseases associated with advancing age.

Cell Senescence

The phenomenon of cell senescence was described for the first time by Leonard Hayflick and Paul Moorhead in 1961 using diploid fibroblast cell lines, that were unable to divide after 40–60 population doublings [35]. Cell senescence indicates the prolonged and generally irreversible arrest of cell proliferation that is induced by a large number of cellular stressors [36, 37]. These include DNA damage, telomere shortening, radiation, reactive oxygen species (ROS) and other reactive metabolites, mitogenic and metabolic stressors [38, 39]. It has also been postulated that chronic viral infections, gut dysbiosis, pollution and industrial toxicants may be involved [40]. Genotoxic and cytotoxic drugs widely used as anti-cancer chemotherapeutics also induce senescence and cause long-term side effects similar to those observed in individuals suffering from age-associated pathologies [41, 42]. Different stressors give rise to different types of senescence, such as telomere-dependent replicative senescence, non telomeric stress-induced premature senescence, DNA damage-induced senescence, epigenetically-induced cell senescence and mitochondrial dysfunction-associated cell senescence, as reviewed in [43].

There is not a universal marker to identify senescent cells, but $p16^{INK4a}$ and $p21^{CIP1/WAF1}$ (the cyclin-dependent kinase inhibitors 2A and 1A, CDKN2A and CDKN1A, respectively) are widely used to visualize them both in vivo and in vitro [44]. Senescent cells can also express senescence-associated β-galactosidase (SA-β-gal) [44], the enzyme that catalyzes the hydrolysis of β-galactosides into monosaccharides. SA β-gal detected at pH 6.0 allows the identification of senescent cells in culture and in tissues. The markers $p16^{INK4a}$ and $p21^{CIP1/WAF1}$ are not always co-expressed in tissued from aging mice and humans, with some organs (pancreas, liver, intestine, spleen and brain cortex) showing increased expression of $p16^{INK4a}$ with age, whereas the epidermis shows increased expression of $p21^{CIP1/WAF1}$, suggesting that different tissues trigger different senescent programs [45].

Cell senescence was initially thought to be a tissue culture artifact, but many studies have clearly shown that it also occurs in vivo, and it takes part in many different physiological and pathological processes [46]. Proliferation arrest is mediated by inhibition of cell cycle progression through $p16^{INK4a}$ and/or the activation of cell cycle arrest through $p53/p21^{CIP1/WAF1}$. Mechanistically, induction of $p21^{CIP1/WAF1}$ and $p16^{INK4a}$ inhibits cyclin-dependent kinases, necessary to promote cell cycle progression. Despite experiencing irreversible growth arrest, senescent cells are metabolically and transcriptionally active, undergo characteristic morphological, transcriptional and epigenetic changes, and changes in protein expression and secretion, developing a complex secretome known as the senescence-associated secretory phenotype,

SASP. The SASP represents the major communication tool between senescent cells and the microenvironment, although other ways of communication have been identified. These include direct interaction through cell–cell contact [47], cell fusion [48], formation of cytoplasmic bridges [49], and extracellular vesicles [50].

Cells that have been shown to undergo senescence and secrete products of the SASP are endothelial and epithelial cells, liver stellate cells [51], as well as immune cells [34, 52, 53]. In addition, it was found that muscle cells from women aged from 65 to 71 years can express $p21^{CIP1/WAF1}$ at higher levels than those from young women. These findings suggest that aging-associated impairment of muscle function, regeneration and volume occurs via SASP [54].

The SASP

Heterogeneous Composition of the SASP

The cell type undergoing senescence and the stimulus inducing senescence determine the composition of the SASP. The SASP is responsible for the secretion of several soluble factors, such as pro-inflammatory cytokines (IL-1α, IL-1β, IL-6, TNF-α, IFN-γ), chemokines (IL-8, MCP-2, MCP-4, MIP1α, Eotaxin), soluble receptors (TNF receptors), non-protein soluble factors (nitric oxide, ROS), growth factors (Epidermal Growth Factor, Fibroblast Growth Factor, Nerve Growth Factor, Vascular Endothelial Growth Factor), extracellular matrix macromolecules (Fibronectin, Collagens, Laminin), growth modulators (amphiregulin, epiregulin, heregulin), pro-survival factors (neurotrophic factor), proteases and regulators), hemostatic factors, ceramides, bradykinins, damage-associated molecular patterns, DAMPs [39, 51], and also the B cell-activating factor BAFF as recently shown [55].

Transcriptional and Post-Transcriptional Regulation

Transcriptionally, the SASP is regulated by the activation of two major transcription factors: NF-κB and CCAAT/enhancer binding protein β (C/EBPβ) [56], and loss of these transcription factors has been shown to significantly reduce the secretion of pro-inflammatory cytokines and chemokines.

The SASP is also regulated post-transcriptionally through activation of the mammalian target of rapamycin (mTOR), and through the translation of mitogen-activated protein kinase activated protein kinase 2 (MAPKAPK2), downstream from p38, a member of the mitogen-activated protein kinase (MAPK) family, which phosphorylates and inhibits the RNA-binding protein ZFP36L1 during senescence, inhibiting its ability to degrade the transcripts of several SASP components [57]. In contrast, inhibition of mTOR has been shown to prevent the induction of the

SASP. The polypyrimidine tract-binding protein 1 (PTBP1) has also been described to regulate alternative splicing of genes involved in intracellular trafficking, such as EXOC7, and control SASP expression [58].

Intracellular Signaling Pathways

Signaling pathways that regulate the SASP include activation of p38, leading to NF-κB translocation into the nucleus and activation [59]; metabolic fluctuations and altered nicotinamide adenine dinucleotide, NAD+/NADH ratios, leading to the 5′AMP-activated protein kinase (AMPK)-mediated activation of p53 [60]; activation of the cyclic GMP-AMP synthase (cGAS)-stimulator of Interferon Genes (STING), in response to the accumulation of cytoplasmic DNA (cytoplasmic chromatin fragments and mitochondrial) in senescent cells [61].

Epigenetic Regulation

Chromatin modifications are also involved in the regulation of the SASP. These modifications include histone acetylation by acetyltransferases and methylation of histone lysine and arginine residues. The senescence-associated decrease in the expression of SIRT1, a NAD+-dependent protein deacetylase, allows the transcription of SASP markers through increased acetylation of the H3K9 and H4K16 histones, as shown in the promoter regions of the IL-6 and IL-8 genes [62]. The decreased demethylation of histone H3K9 at promoter regions of several SASP factors, leads to their up-regulated transcription [63], Moreover, increased expression of the histone variant macroH2A1 accumulates in senescent cells [64]. The high-mobility-group protein B2 (HMGB2), involved in chromatin remodeling and transcription regulation, also regulates the SASP by localizing to SASP loci and facilitating transcription of several target genes, and HMGB2 inhibition in senescent cells blocks neoplastic transformation and expression of pro-inflammatory and pro-tumorigenic effects of SASP factors [65].

Pleiotropic Effects of the SASP

The SASP has pleiotropic, sometimes opposite, paradoxical effects. The reasons rely on the type of stimulus that triggers senescence and determines the heterogeneous, dynamic and complex composition of the SASP, as well as on the context of the tissue microenvironment, on the cell type undergoing senescence, and on the stage of senescence.

Senescent cells play beneficial roles in many physiological and pathological processes and are involved in embryonic morphogenesis [66], wound healing [67], restoration of tissue homeostasis in chronic liver fibrosis [68] and tumor suppression [69, 70].

Senescence occurs during mammalian embryonic development at multiple locations. Mechanistically, senescence is $p21^{CIP1/WAF1}$-dependent but DNA damage- or p53-independent, and is regulated by TGF-β/SMAD and PI3K/FOXO signaling pathways, followed by macrophage infiltration, clearance of senescent cells, and tissue remodeling. These findings demonstrate that developmentally-programmed senescence promotes tissue remodeling [66].

In wound healing, senescent fibroblasts and endothelial cells appear early in response to cutaneous wounds, where they accelerate wound closure by secreting platelet-derived growth factor AA (PDGF-AA), a factor that promotes myofibroblast differentiation [67]. To support these findings, mouse models in which a senescence-sensitive promoter in the tumor suppressor $p16^{INK4a}$ has been used to track senescent cells in vivo, have shown that topical treatment of senescence-free wounds with recombinant PDGF-AA was able to rescue delayed wound closure and lack of myofibroblast differentiation.

In the presence of chronic liver damage due to chronic hepatitis- or alcohol-induced damage, tumor initiation is often preceded by hepatic fibrosis. It has been shown that ablation of a p53-dependent senescence program in hepatic stellate cells increases liver fibrosis and cirrhosis associated with reduced survival and enhances the transformation of adjacent epithelial cells into hepatocellular carcinoma (HCC). Mechanistically, p53-expressing senescent stellate cells release factors that activate macrophage polarization toward a tumor-inhibiting M1 phenotype capable of attacking senescent cells in culture, whereas proliferating p53-deficient stellate cells secrete factors that stimulate polarization of macrophages into a tumor-promoting M2 phenotype and enhance the proliferation of premalignant cells. Therefore, p53 can act to suppress tumorigenesis by promoting an anti-tumor microenvironment, through secreted factors that modulate macrophage function [68].

In cancer, senescence-associated cell-cycle exit, evolved as an anti-tumor mechanism, prevents cancer cells from proliferation. Senescence can also be beneficial for cancer therapy by re-engaging anti-tumor immunity in immunologically "cold" tumors, such as pancreatic cancer, through its non-cell autonomous effects on the tumor microenvironment. SASP induction following therapy-induced senescence can recruit and activate cytotoxic lymphocytes such as NK cells and CD8+ T cells in lung and pancreatic tumors, leading to anti-tumor immunity, disease regression, and synergy with checkpoint blockade immunotherapy [69, 70].

All the beneficial roles of senescent cells seem to occur in young mice and humans. However, in older individuals, senescent cells accumulate in the body with adverse consequences, leading to the development of several age-associated conditions and diseases, such as T2DM, cancer, neurodegenerative diseases and cardiovascular disease [71, 72]. Secretion of SASP products leads to cell death, tissue remodeling and infiltration of pro-inflammatory immune cells in tissues that contribute to local and systemic inflammation [73, 74]. SASP factors can have immunosuppressive effects and contribute to tumor surveillance and tumor progression, as shown for example in HCCs where pre-neoplastic hepatocytes secrete the chemokine CCL2, part of the SASP [75], which promotes the recruitment of immature myeloid cells that inhibit NK cells with cytotoxic function [76]. Moreover, the panel of proteases secreted by senescent cells, such as serine proteases and regulators of the plasminogen activation pathway, overlaps in large part with those found in malignant tumors [77].

Cell Senescence and the SASP Contribute to Aging and Age-Associated Diseases

Whether and how cell senescence contributes to aging, age-associated conditions and diseases, and to frailty and physical dysfunction is one of the major open questions in the biology of aging field [78]. It is known that the loss of cells capable to proliferate may be responsible for age-associated pathologies like glaucoma and cataracts [79, 80], type 1 diabetes [81] and osteoarthritis [82]. It is also known that inflammatory SASP products may play a role in age-associated diseases like atherosclerosis [83] and cancer [84], whereas extracellular matrix macromolecules may regulate pulmonary fibrosis [85].

SASP markers are present in plasma samples of elderly individuals [86]. The "SASP Atlas" (www.SASPAtlas.com), a comprehensive proteomic database of soluble SASP factors derived from multiple senescence inducers and different cell types, has allowed the identification of several candidate biomarkers of cellular senescence that overlap with aging markers in human plasma, including markers that significantly correlated with age in plasma samples from the human Baltimore Longitudinal Study of Aging. Although the SASP contributes to systemic low-grade chronic inflammation, inflammaging, specific SASP factors secreted by senescent cells might also be responsible for local tissue dysfunction.

The transplant of small numbers of senescent pre-adipocytes into young mice has been shown to be sufficient to induce physical dysfunction, such as reduced maximal speed, hanging endurance and grip strength [87]. To test the contribution of the SASP to the induction of cell senescence in vivo, the authors of the paper above analyzed IL-10 knockout mice. These mice have a genetically induced premature pro-inflammatory phenotype that resembles the SASP and also prematurely develop physical dysfunction similar to human frailty [88]. Results showed that these mice have more senescent cells than controls, clearly indicating that the SASP is a crucial mediator of the detrimental effects of senescent cells in vivo.

Using reporter mice with p16^{INK4a}-expressing senescent cells, it has been shown that senescent cells progressively increase with age and drive aging and cancer processes in mice [89, 90]. Moreover, in mouse models of accelerated aging, genomic instability leading to accelerated accumulation of senescent cells results in symptoms of premature aging, such as shortened lifespan and accelerated onset of age-related diseases [91].

Proof-of-principle experiments conducted in mice have shown that removal of senescent cells increases life span [79]. Briefly, mice expressing the transgene *INK-ATTAC* for inducible elimination of $p16^{Ink4a}$-positive senescent cells upon administration of a drug, showed reduced amounts of senescent cells in the adipose tissue, in the skeletal muscle and in the eye, tissues in which $p16^{Ink4a}$ contributes to the acquisition of age-related pathologies. In these mice, life-long removal of cells expressing $p16^{Ink4a}$ delayed onset of these phenotypes, whereas senescent cell clearance later in life attenuated progression of already established age-related disorders.

The accumulation of senescent cells in tissues often leads to the depletion of functional cells and to the physiological decline in progenitor cells, needed for tissue homeostasis, repair and regeneration. The frequency, proliferation and neurogenesis of stem cells and other progenitors in the subventricular zone of the olfactory bulb was tested in young and old $p16^{Ink4a}$-deficient mice, and of littermate controls. Results showed that deletion of $p16^{INK4a}$ in young mice did not significantly affect the frequency of cells that formed multipotent neurospheres in culture, as well as their self-renewal. Conversely, deletion of $p16^{INK4a}$ in old mice significantly increased the frequency of cells that formed multipotent neurospheres in culture, suggesting that increased $p16^{INK4a}$ expression with age is one of the factors responsible for age-driven reduced progenitor function and neurogenesis observed in the nervous system [92].

Cell senescence also contributes to diseases through the secretion of SASP markers involved in tissue remodeling. An example is idiopathic pulmonary fibrosis, a fatal disease characterized by interstitial remodeling that leads to disruption of tissue homeostasis and compromised lung function [85]. Senescence biomarkers accumulate in the damaged lung, as well as senescent fibroblasts characterized by a pro-fibrotic secretome. Senescent cell retention in the damaged lung is associated with impaired ability to resolve injury. Moreover, senescent cell elimination has been shown to significantly improve health outcomes but only when treatment is done at an early-stage pathogenesis.

Similarly, the accumulation of senescent foamy macrophages in the sub-endothelial space at the onset of atherosclerosis, leads to increased secretion of inflammatory cytokines/chemokines and atherogenic markers, and contributes to pathologic outcomes, promoting features of plaque instability, including elastic fiber degradation and fibrous cap thinning, by heightening metalloprotease production [93]. These senescent cells that accumulate in the atherosclerotic lesions express high levels of SASP transcripts such as $p16^{INK4a}$, $p19^{Arf}$, the matrix metalloproteases Mmp3 and Mmp13, and the inflammatory cytokines IL-1α and TNF-α. Clearing senescent cells from advanced lesions inhibits maladaptive plaque remodeling processes associated with plaque rupture, suggesting that senescent cells can directly influence pro-atherogenic processes through specific SASP factors.

Targeting Senescent Cells and the SASP

Senescent cells and the SASP are clearly key therapeutic targets for aging and age-associated conditions and diseases. Senescent cells can be effectively targeted using genetic approaches, as those utilized in the examples above, and/or senolytics, as well as with small molecules that block the SASP and its detrimental effects (senomorphics), leading to significant improvement of tissue function in different in vivo models of aging and of age-associated diseases [94, 95], thus suggesting a potential clinical use. Experiments in mice have indeed shown that the clearance of $p16^{INK4a}$-positive cells delays or prevents several age-associated pathologies such as sarcopenia, cataracts, cardiac hypertrophy, kidney disease, cancer, atherosclerosis, osteoarthritis, and neurodegeneration [45].

Research to discover senolytic drugs started in the early 2000s, when the group of Norman Sharpless showed that the accumulation of senescent cells, identified by positive $p16^{INK4a}$ and SA-β-gal staining, was delayed in mice calorically restricted [96]. Then in 2005 the group of James Kirkland applied high-throughput screening approaches and other traditional strategies for drug development to discover senolytics, based on previously published findings that senescent cells are highly resistant to cell death as they have up-regulated anti-apoptotic, pro-survival pathways [97], and prevent their own removal from the tissues using senescent cell anti-apoptotic protective (SCAP) networks [98]. SCAP networks were found to be highly expressed in senescent cells but not in non-senescent cells. It was shown that SASP-expressing senescent cells rely on SCAP networks to escape cell death and survive [98]. Kirkland's group then used bioinformatics approaches to discover molecules able to target SCAP networks and identified almost 50 potential senolytic drugs. Therefore, the first senolytics were identified among natural products already in use in clinical studies, mostly as anti-cancer drugs, to facilitate translation from benchside to clinical research. These compounds included Dasatinib (D), a tyrosine kinase inhibitor approved in the US in 2006 for clinical use [99, 100], and two flavonoids, Quercetin (Q) and Fisetin (F) [101]. These flavonoids act by inhibiting members of the BCL-2 family and other components of SCAP networks [102, 103]. In many cases, the effects of 2 drugs together was better than each one alone, as shown by the senolytic effects of D + Q on human senescent HUVECs and preadipocytes [98]. Research on the development of senolytics has clearly shown that the best senolytics are those that target multiple SCAP members and networks, whereas drugs that act blocking a single receptor or enzyme, or a single signaling pathway, have significant off-target effects and induce apoptosis also in non-senescent cells, including immune cells [104, 105]. Also, the target of senolytics is the senescent cell and not a single product or pathway that can be targeted with senomorphics. Conversely, the combination D + Q and F have been shown to significantly decrease the frequency of senescent cells in vivo without having any effect on non-senescent cells [87, 106]. Several clinical trials with senolytics are currently underway and many others have already been planned, and multicenter collaborations have been created in order to conduct interventional trials that target fundamental processes of aging as those in the NIH-funded Translational Geroscience Network and the Alliance for Healthy Aging.

There is a dynamic relationship between senescence and nutrition, and micronutrients (Vitamin D, Vitamin E, Zinc), as well as dietary patterns (Mediterranean diet, caloric restriction), have been suggested as good strategies to counteract the effects of age-associated alterations in immune cells that lead to age-associated pathologies [107–112]. In addition to Q and F described above, other nutritional senolytics have been discovered. Among these, the mostly studied are curcumin and resveratrol. Curcumin is an extract from turmeric (*Curcuma longa*) with anti-inflammatory characteristics, i.e. reduction of NF-kB activation and consequent secretion of the pro-inflammatory cytokines IL-1β and TNF-α, while enhancing IL-10 secretion [113, 114]. The beneficial effects of curcumin have been primarily shown in clinical trials with patients suffering of inflammatory diseases, such as cancer, arthritis, uveitis, ulcerative and Chron's diseases, peptic ulcer, acute coronary syndrome [115]. Resveratrol, a flavonoid abundant in red wine and fruits, acts through the activation of SIRT1, metabolic reprogramming of CD4+ T cells, reduction of their glycolytic and pro-inflammatory status and increase in their effector function [116]. The best effects of resveratrol have been observed in patients with pathologies related to oxidative stress, T2DM, cardiovascular and neurological diseases [117].

Conclusive Remarks

With the older population growing at unprecedented rates, the increasing accumulation of senescent cells in body's tissues promotes inflammaging and age-related diseases. Therefore, therapeutic interventions targeting either senescent cells or the SASP represent promising strategies for promoting healthy aging. Although ongoing clinical trials have shown that senotherapeutic approaches could be beneficial in several human diseases, larger clinical trials are still needed to translate the results into patient care, allowing the development of individually tailored treatments.

Compliance With Ethical Standards

Conflict of Interest The author declares that she has not conflict of interest.

References

1. Franceschi C, Campisi J. Chronic inflammation (inflammaging) and its potential contribution to age-associated diseases. J Gerontol A Biol Sci Med Sci. 2014;69 Suppl 1:S4-9. https://doi.org/10.1093/gerona/glu057.
2. Franceschi C, Bonafe M, Valensin S, Olivieri F, De Luca M, Ottaviani E, et al. Inflamm-aging. An evolutionary perspective on immunosenescence. Ann N Y Acad Sci. 2000;908:244–54. https://doi.org/10.1111/j.1749-6632.2000.tb06651.x.

3. Serrano-Lopez J, Martin-Antonio B. Inflammaging, an Imbalanced Immune Response That Needs to Be Restored for Cancer Prevention and Treatment in the Elderly. Cells. 2021;10(10). https://doi.org/10.3390/cells10102562.
4. Zinger A, Cho WC, Ben-Yehuda A. Cancer and Aging - the Inflammatory Connection. Aging Dis. 2017;8(5):611–27. https://doi.org/10.14336/AD.2016.1230.
5. Lindholm E, Bakhtadze E, Cilio C, Agardh E, Groop L, Agardh CD. Association between LTA, TNF and AGER polymorphisms and late diabetic complications. PLoS One. 2008;3(6):e2546. https://doi.org/10.1371/journal.pone.0002546.
6. Sarzi-Puttini P, Atzeni F, Doria A, Iaccarino L, Turiel M. Tumor necrosis factor-alpha, biologic agents and cardiovascular risk. Lupus. 2005;14(9):780–4. https://doi.org/10.1191/0961203305lu2220oa.
7. Holmes C, Cunningham C, Zotova E, Woolford J, Dean C, Kerr S, et al. Systemic inflammation and disease progression in Alzheimer disease. Neurology. 2009;73(10):768–74. https://doi.org/10.1212/WNL.0b013e3181b6bb95.
8. Isaacs JD. Therapeutic agents for patients with rheumatoid arthritis and an inadequate response to tumour necrosis factor-alpha antagonists. Expert Opin Biol Ther. 2009;9(12):1463–75. https://doi.org/10.1517/14712590903379494.
9. Calabrese V, Santoro A, Monti D, Crupi R, Di Paola R, Latteri S, et al. Aging and Parkinson's Disease: Inflammaging, neuroinflammation and biological remodeling as key factors in pathogenesis. Free Radic Biol Med. 2018;115:80–91. https://doi.org/10.1016/j.freeradbiomed.2017.10.379.
10. Bolton C. An evaluation of the recognised systemic inflammatory biomarkers of chronic sub-optimal inflammation provides evidence for inflammageing (IFA) during multiple sclerosis (MS). Immun Ageing. 2021;18(1):18. https://doi.org/10.1186/s12979-021-00225-0.
11. He Y, Chen Y, Yao L, Wang J, Sha X, Wang Y. The Inflamm-Aging Model Identifies Key Risk Factors in Atherosclerosis. Front Genet. 2022;13:865827. https://doi.org/10.3389/fgene.2022.865827.
12. Tan W, Zou J, Yoshida S, Jiang B, Zhou Y. The Role of Inflammation in Age-Related Macular Degeneration. Int J Biol Sci. 2020;16(15):2989–3001. https://doi.org/10.7150/ijbs.49890.
13. Mundy GR. Osteoporosis and inflammation. Nutr Rev. 2007;65(12 Pt 2):S147–51. https://doi.org/10.1111/j.1753-4887.2007.tb00353.x.
14. Alberro A, Iribarren-Lopez A, Saenz-Cuesta M, Matheu A, Vergara I, Otaegui D. Inflammaging markers characteristic of advanced age show similar levels with frailty and dependency. Sci Rep. 2021;11(1):4358. https://doi.org/10.1038/s41598-021-83991-7.
15. Verschoor CP, Belsky DW, Andrew MK, Haynes L, Loeb M, Pawelec G, et al. Advanced biological age is associated with improved antibody responses in older high-dose influenza vaccine recipients over four consecutive seasons. Immun Ageing. 2022;19(1):39. https://doi.org/10.1186/s12979-022-00296-7.
16. Giacconi R, Cipriano C, Albanese F, Boccoli G, Saba V, Olivieri F, et al. The -174G/C polymorphism of IL-6 is useful to screen old subjects at risk for atherosclerosis or to reach successful ageing. Exp Gerontol. 2004;39(4):621–8. https://doi.org/10.1016/j.exger.2003.12.013.
17. Lio D, Marino V, Serauto A, Gioia V, Scola L, Crivello A, et al. Genotype frequencies of the + 874T-->A single nucleotide polymorphism in the first intron of the interferon-gamma gene in a sample of Sicilian patients affected by tuberculosis. Eur J Immunogenet. 2002;29(5):371–4. https://doi.org/10.1046/j.1365-2370.2002.00327.x.
18. Olivieri F, Bonafe M, Cavallone L, Giovagnetti S, Marchegiani F, Cardelli M, et al. The -174 C/G locus affects in vitro/in vivo IL-6 production during aging. Exp Gerontol. 2002;37(2–3):309–14. https://doi.org/10.1016/s0531-5565(01)00197-8.
19. Pravica V, Perrey C, Stevens A, Lee JH, Hutchinson IV. A single nucleotide polymorphism in the first intron of the human IFN-gamma gene: absolute correlation with a polymorphic CA microsatellite marker of high IFN-gamma production. Hum Immunol. 2000;61(9):863–6. https://doi.org/10.1016/s0198-8859(00)00167-1.
20. Lio D, Scola L, Crivello A, Colonna-Romano G, Candore G, Bonafe M, et al. Gender-specific association between -1082 IL-10 promoter polymorphism and longevity. Genes Immun. 2002;3(1):30–3. https://doi.org/10.1038/sj.gene.6363827.

21. Freeman RB, Jr. The 'indirect' effects of cytomegalovirus infection. Am J Transplant. 2009;9(11):2453–8. https://doi.org/10.1111/j.1600-6143.2009.02824.x.
22. Haspot F, Lavault A, Sinzger C, Laib Sampaio K, Stierhof YD, Pilet P, et al. Human cytomegalovirus entry into dendritic cells occurs via a macropinocytosis-like pathway in a pH-independent and cholesterol-dependent manner. PLoS One. 2012;7(4):e34795. https://doi.org/10.1371/journal.pone.0034795.
23. Heintz C, Mair W. You are what you host: microbiome modulation of the aging process. Cell. 2014;156(3):408–11. https://doi.org/10.1016/j.cell.2014.01.025.
24. Pinti M, Cevenini E, Nasi M, De Biasi S, Salvioli S, Monti D, et al. Circulating mitochondrial DNA increases with age and is a familiar trait: Implications for "inflamm-aging". Eur J Immunol. 2014;44(5):1552–62. https://doi.org/10.1002/eji.201343921.
25. Zhang Q, Raoof M, Chen Y, Sumi Y, Sursal T, Junger W, et al. Circulating mitochondrial DAMPs cause inflammatory responses to injury. Nature. 2010;464(7285):104–7. https://doi.org/10.1038/nature08780.
26. Meier J, Sturm A. The intestinal epithelial barrier: does it become impaired with age? Dig Dis. 2009;27(3):240–5. https://doi.org/10.1159/000228556.
27. Forsythe LK, Wallace JM, Livingstone MB. Obesity and inflammation: the effects of weight loss. Nutr Res Rev. 2008;21(2):117–33. https://doi.org/10.1017/S0954422408138732.
28. Ponti F, Santoro A, Mercatelli D, Gasperini C, Conte M, Martucci M, et al. Aging and Imaging Assessment of Body Composition: From Fat to Facts. Front Endocrinol (Lausanne). 2019;10:861. https://doi.org/10.3389/fendo.2019.00861.
29. Tchernof A, Despres JP. Pathophysiology of human visceral obesity: an update. Physiol Rev. 2013;93(1):359–404. https://doi.org/10.1152/physrev.00033.2011.
30. Tchkonia T, Morbeck DE, Von Zglinicki T, Van Deursen J, Lustgarten J, Scrable H, et al. Fat tissue, aging, and cellular senescence. Aging Cell. 2010;9(5):667–84. https://doi.org/10.1111/j.1474-9726.2010.00608.x.
31. Olivieri F, Rippo MR, Procopio AD, Fazioli F. Circulating inflamma-miRs in aging and age-related diseases. Front Genet. 2013;4:121. https://doi.org/10.3389/fgene.2013.00121.
32. Ovadya Y, Landsberger T, Leins H, Vadai E, Gal H, Biran A, et al. Impaired immune surveillance accelerates accumulation of senescent cells and aging. Nat Commun. 2018;9(1):5435. https://doi.org/10.1038/s41467-018-07825-3.
33. Freund A, Orjalo AV, Desprez PY, Campisi J. Inflammatory networks during cellular senescence: causes and consequences. Trends Mol Med. 2010;16(5):238–46. https://doi.org/10.1016/j.molmed.2010.03.003.
34. Sikora E, Arendt T, Bennett M, Narita M. Impact of cellular senescence signature on ageing research. Ageing Res Rev. 2011;10(1):146–52. https://doi.org/10.1016/j.arr.2010.10.002.
35. Hayflick L, Moorhead PS. The serial cultivation of human diploid cell strains. Exp Cell Res. 1961;25:585–621. https://doi.org/10.1016/0014-4827(61)90192-6.
36. Hayflick L. The Limited in Vitro Lifetime of Human Diploid Cell Strains. Exp Cell Res. 1965;37:614–36. https://doi.org/10.1016/0014-4827(65)90211-9.
37. Munoz-Espin D, Serrano M. Cellular senescence: from physiology to pathology. Nat Rev Mol Cell Biol. 2014;15(7):482–96. https://doi.org/10.1038/nrm3823.
38. Birch J, Gil J. Senescence and the SASP: many therapeutic avenues. Genes Dev. 2020;34(23-24):1565–76. https://doi.org/10.1101/gad.343129.120.
39. Campisi J. Cellular senescence: putting the paradoxes in perspective. Curr Opin Genet Dev. 2011;21(1):107–12. https://doi.org/10.1016/j.gde.2010.10.005.
40. Furman D, Campisi J, Verdin E, Carrera-Bastos P, Targ S, Franceschi C, et al. Chronic inflammation in the etiology of disease across the life span. Nat Med. 2019;25(12):1822–32. https://doi.org/10.1038/s41591-019-0675-0.
41. Demaria M, O'Leary MN, Chang J, Shao L, Liu S, Alimirah F, et al. Cellular Senescence Promotes Adverse Effects of Chemotherapy and Cancer Relapse. Cancer Discov. 2017;7(2):165–76. https://doi.org/10.1158/2159-8290.CD-16-0241.
42. Sun Y, Campisi J, Higano C, Beer TM, Porter P, Coleman I, et al. Treatment-induced damage to the tumor microenvironment promotes prostate cancer therapy resistance through WNT16B. Nat Med. 2012;18(9):1359–68. https://doi.org/10.1038/nm.2890.

43. Toussaint O, Royer V, Salmon M, Remacle J. Stress-induced premature senescence and tissue ageing. Biochem Pharmacol. 2002;64(5–6):1007–9. https://doi.org/10.1016/s0006-2952(02)01170-x.
44. Hernandez-Segura A, Nehme J, Demaria M. Hallmarks of Cellular Senescence. Trends Cell Biol. 2018;28(6):436–53. https://doi.org/10.1016/j.tcb.2018.02.001.
45. Idda ML, McClusky WG, Lodde V, Munk R, Abdelmohsen K, Rossi M, et al. Survey of senescent cell markers with age in human tissues. Aging (Albany NY). 2020;12(5):4052–66. https://doi.org/10.18632/aging.102903.
46. Burton DG, Krizhanovsky V. Physiological and pathological consequences of cellular senescence. Cell Mol Life Sci. 2014;71(22):4373–86. https://doi.org/10.1007/s00018-014-1691-3.
47. Nelson G, Wordsworth J, Wang C, Jurk D, Lawless C, Martin-Ruiz C, et al. A senescent cell bystander effect: senescence-induced senescence. Aging Cell. 2012;11(2):345–9. https://doi.org/10.1111/j.1474-9726.2012.00795.x.
48. Chuprin A, Gal H, Biron-Shental T, Biran A, Amiel A, Rozenblatt S, et al. Cell fusion induced by ERVWE1 or measles virus causes cellular senescence. Genes Dev. 2013;27(21):2356–66. https://doi.org/10.1101/gad.227512.113.
49. Biran A, Zada L, Abou Karam P, Vadai E, Roitman L, Ovadya Y, et al. Quantitative identification of senescent cells in aging and disease. Aging Cell. 2017;16(4):661–71. https://doi.org/10.1111/acel.12592.
50. Takasugi M, Okada R, Takahashi A, Virya Chen D, Watanabe S, Hara E. Small extracellular vesicles secreted from senescent cells promote cancer cell proliferation through EphA2. Nat Commun. 2017;8:15729. https://doi.org/10.1038/ncomms15728.
51. Coppe JP, Desprez PY, Krtolica A, Campisi J. The senescence-associated secretory phenotype: the dark side of tumor suppression. Annu Rev Pathol. 2010;5:99–118. https://doi.org/10.1146/annurev-pathol-121808-102144.
52. Frasca D, Diaz A, Romero M, Blomberg BB. Human peripheral late/exhausted memory B cells express a senescent-associated secretory phenotype and preferentially utilize metabolic signaling pathways. Exp Gerontol. 2017;87(Pt A):113–20. https://doi.org/10.1016/j.exger.2016.12.001.
53. Frasca D, Romero M, Diaz A, Garcia D, Thaller S, Blomberg BB. B Cells with a Senescent-Associated Secretory Phenotype Accumulate in the Adipose Tissue of Individuals with Obesity. Int J Mol Sci. 2021;22(4). https://doi.org/10.3390/ijms22041839.
54. Welle S, Brooks AI, Delehanty JM, Needler N, Bhatt K, Shah B, et al. Skeletal muscle gene expression profiles in 20-29 year old and 65-71 year old women. Exp Gerontol. 2004;39(3):369-77. https://doi.org/10.1016/j.exger.2003.11.011.
55. Rossi M, Anerillas C, Idda ML, Munk R, Shin CH, Donega S, et al. Pleiotropic effects of BAFF on the senescence-associated secretome and growth arrest. Elife. 2023;12. https://doi.org/10.7554/eLife.84238.
56. Kuilman T, Michaloglou C, Vredeveld LC, Douma S, van Doorn R, Desmet CJ, et al. Oncogene-induced senescence relayed by an interleukin-dependent inflammatory network. Cell. 2008;133(6):1019–31. https://doi.org/10.1016/j.cell.2008.03.039.
57. Herranz N, Gallage S, Mellone M, Wuestefeld T, Klotz S, Hanley CJ, et al. mTOR regulates MAPKAPK2 translation to control the senescence-associated secretory phenotype. Nat Cell Biol. 2015;17(9):1205–17. https://doi.org/10.1038/ncb3225.
58. Georgilis A, Klotz S, Hanley CJ, Herranz N, Weirich B, Morancho B, et al. PTBP1-Mediated Alternative Splicing Regulates the Inflammatory Secretome and the Pro-tumorigenic Effects of Senescent Cells. Cancer Cell. 2018;34(1):85–102 e9. https://doi.org/10.1016/j.ccell.2018.06.007.
59. Freund A, Patil CK, Campisi J. p38MAPK is a novel DNA damage response-independent regulator of the senescence-associated secretory phenotype. EMBO J. 2011;30(8):1536–48. https://doi.org/10.1038/emboj.2011.69.
60. Wiley CD, Velarde MC, Lecot P, Liu S, Sarnoski EA, Freund A, et al. Mitochondrial Dysfunction Induces Senescence with a Distinct Secretory Phenotype. Cell Metab. 2016;23(2):303–14. https://doi.org/10.1016/j.cmet.2015.11.011.

61. Loo TM, Miyata K, Tanaka Y, Takahashi A. Cellular senescence and senescence-associated secretory phenotype via the cGAS-STING signaling pathway in cancer. Cancer Sci. 2020;111(2):304–11. https://doi.org/10.1111/cas.14266.
62. Hayakawa T, Iwai M, Aoki S, Takimoto K, Maruyama M, Maruyama W, et al. SIRT1 suppresses the senescence-associated secretory phenotype through epigenetic gene regulation. PLoS One. 2015;10(1):e0116480. https://doi.org/10.1371/journal.pone.0116480.
63. Takahashi A, Imai Y, Yamakoshi K, Kuninaka S, Ohtani N, Yoshimoto S, et al. DNA damage signaling triggers degradation of histone methyltransferases through APC/C(Cdh1) in senescent cells. Mol Cell. 2012;45(1):123–31. https://doi.org/10.1016/j.molcel.2011.10.018.
64. Chen H, Ruiz PD, McKimpson WM, Novikov L, Kitsis RN, Gamble MJ. MacroH2A1 and ATM Play Opposing Roles in Paracrine Senescence and the Senescence-Associated Secretory Phenotype. Mol Cell. 2015;59(5):719–31. https://doi.org/10.1016/j.molcel.2015.07.011.
65. Aird KM, Iwasaki O, Kossenkov AV, Tanizawa H, Fatkhutdinov N, Bitler BG, et al. HMGB2 orchestrates the chromatin landscape of senescence-associated secretory phenotype gene loci. J Cell Biol. 2016;215(3):325–34. https://doi.org/10.1083/jcb.201608026.
66. Munoz-Espin D, Canamero M, Maraver A, Gomez-Lopez G, Contreras J, Murillo-Cuesta S, et al. Programmed cell senescence during mammalian embryonic development. Cell. 2013;155(5):1104–18. https://doi.org/10.1016/j.cell.2013.10.019.
67. Demaria M, Ohtani N, Youssef SA, Rodier F, Toussaint W, Mitchell JR, et al. An essential role for senescent cells in optimal wound healing through secretion of PDGF-AA. Dev Cell. 2014;31(6):722–33. https://doi.org/10.1016/j.devcel.2014.11.012.
68. Lujambio A, Akkari L, Simon J, Grace D, Tschaharganeh DF, Bolden JE, et al. Non-cell-autonomous tumor suppression by p53. Cell. 2013;153(2):449–60. https://doi.org/10.1016/j.cell.2013.03.020.
69. Ruscetti M, Leibold J, Bott MJ, Fennell M, Kulick A, Salgado NR, et al. NK cell-mediated cytotoxicity contributes to tumor control by a cytostatic drug combination. Science. 2018;362(6421):1416–22. https://doi.org/10.1126/science.aas9090.
70. Ruscetti M, Morris JPt, Mezzadra R, Russell J, Leibold J, Romesser PB, et al. Senescence-Induced Vascular Remodeling Creates Therapeutic Vulnerabilities in Pancreas Cancer. Cell. 2020;181(2):424–41 e21. https://doi.org/10.1016/j.cell.2020.03.008.
71. Campisi J. Aging, cellular senescence, and cancer. Annu Rev Physiol. 2013;75:685–705. https://doi.org/10.1146/annurev-physiol-030212-183653.
72. He S, Sharpless NE. Senescence in Health and Disease. Cell. 2017;169(6):1000–11. https://doi.org/10.1016/j.cell.2017.05.015.
73. Campisi J. Cellular senescence and apoptosis: how cellular responses might influence aging phenotypes. Exp Gerontol. 2003;38(1–2):5–11. https://doi.org/10.1016/s0531-5565(02)00152-3.
74. Passos JF, Nelson G, Wang C, Richter T, Simillion C, Proctor CJ, et al. Feedback between p21 and reactive oxygen production is necessary for cell senescence. Mol Syst Biol. 2010;6:347. https://doi.org/10.1038/msb.2010.5.
75. Acosta JC, Banito A, Wuestefeld T, Georgilis A, Janich P, Morton JP, et al. A complex secretory program orchestrated by the inflammasome controls paracrine senescence. Nat Cell Biol. 2013;15(8):978–90. https://doi.org/10.1038/ncb2784.
76. Eggert T, Wolter K, Ji J, Ma C, Yevsa T, Klotz S, et al. Distinct Functions of Senescence-Associated Immune Responses in Liver Tumor Surveillance and Tumor Progression. Cancer Cell. 2016;30(4):533–47. https://doi.org/10.1016/j.ccell.2016.09.003.
77. Blasi F, Carmeliet P. uPAR: a versatile signalling orchestrator. Nat Rev Mol Cell Biol. 2002;3(12):932–43. https://doi.org/10.1038/nrm977.
78. Childs BG, Durik M, Baker DJ, van Deursen JM. Cellular senescence in aging and age-related disease: from mechanisms to therapy. Nat Med. 2015;21(12):1424–35. https://doi.org/10.1038/nm.4000.
79. Baker DJ, Wijshake T, Tchkonia T, LeBrasseur NK, Childs BG, van de Sluis B, et al. Clearance of p16Ink4a-positive senescent cells delays ageing-associated disorders. Nature. 2011;479(7372):232–6. https://doi.org/10.1038/nature10600.

80. Liton PB, Challa P, Stinnett S, Luna C, Epstein DL, Gonzalez P. Cellular senescence in the glaucomatous outflow pathway. Exp Gerontol. 2005;40(8–9):745–8. https://doi.org/10.1016/j.exger.2005.06.005.
81. Shimizu I, Yoshida Y, Katsuno T, Tateno K, Okada S, Moriya J, et al. p53-induced adipose tissue inflammation is critically involved in the development of insulin resistance in heart failure. Cell Metab. 2012;15(1):51–64. https://doi.org/10.1016/j.cmet.2011.12.006.
82. Martin JA, Brown TD, Heiner AD, Buckwalter JA. Chondrocyte senescence, joint loading and osteoarthritis. Clin Orthop Relat Res. 2004(427 Suppl):S96–103. https://doi.org/10.1097/01.blo.0000143818.74887.b1.
83. Zhou X, Perez F, Han K, Jurivich DA. Clonal senescence alters endothelial ICAM-1 function. Mech Ageing Dev. 2006;127(10):779–85. https://doi.org/10.1016/j.mad.2006.07.003.
84. Thangavel C, Dean JL, Ertel A, Knudsen KE, Aldaz CM, Witkiewicz AK, et al. Therapeutically activating RB: reestablishing cell cycle control in endocrine therapy-resistant breast cancer. Endocr Relat Cancer. 2011;18(3):333–45. https://doi.org/10.1530/ERC-10-0262.
85. Schafer MJ, White TA, Iijima K, Haak AJ, Ligresti G, Atkinson EJ, et al. Cellular senescence mediates fibrotic pulmonary disease. Nat Commun. 2017;8:14532. https://doi.org/10.1038/ncomms14532.
86. Basisty N, Kale A, Jeon OH, Kuehnemann C, Payne T, Rao C, et al. A proteomic atlas of senescence-associated secretomes for aging biomarker development. PLoS Biol. 2020;18(1):e3000599. https://doi.org/10.1371/journal.pbio.3000599.
87. Xu M, Pirtskhalava T, Farr JN, Weigand BM, Palmer AK, Weivoda MM, et al. Senolytics improve physical function and increase lifespan in old age. Nat Med. 2018;24(8):1246–56. https://doi.org/10.1038/s41591-018-0092-9.
88. Walston J, Fedarko N, Yang H, Leng S, Beamer B, Espinoza S, et al. The physical and biological characterization of a frail mouse model. J Gerontol A Biol Sci Med Sci. 2008;63(4):391–8. https://doi.org/10.1093/gerona/63.4.391.
89. Omori S, Wang TW, Johmura Y, Kanai T, Nakano Y, Kido T, et al. Generation of a p16 Reporter Mouse and Its Use to Characterize and Target p16(high) Cells In Vivo. Cell Metab. 2020;32(5):814–28 e6. https://doi.org/10.1016/j.cmet.2020.09.006.
90. Sorrentino JA, Krishnamurthy J, Tilley S, Alb JG, Jr., Burd CE, Sharpless NE. p16INK4a reporter mice reveal age-promoting effects of environmental toxicants. J Clin Invest. 2014;124(1):169–73. https://doi.org/10.1172/JCI70960.
91. Yousefzadeh MJ, Melos KI, Angelini L, Burd CE, Robbins PD, Niedernhofer LJ. Mouse Models of Accelerated Cellular Senescence. Methods Mol Biol. 2019;1896:203–30. https://doi.org/10.1007/978-1-4939-8931-7_17.
92. Molofsky AV, Slutsky SG, Joseph NM, He S, Pardal R, Krishnamurthy J, et al. Increasing p16INK4a expression decreases forebrain progenitors and neurogenesis during ageing. Nature. 2006;443(7110):448–52. https://doi.org/10.1038/nature05091.
93. Childs BG, Baker DJ, Wijshake T, Conover CA, Campisi J, van Deursen JM. Senescent intimal foam cells are deleterious at all stages of atherosclerosis. Science. 2016;354(6311):472–7. https://doi.org/10.1126/science.aaf6659.
94. Myrianthopoulos V, Evangelou K, Vasileiou PVS, Cooks T, Vassilakopoulos TP, Pangalis GA, et al. Senescence and senotherapeutics: a new field in cancer therapy. Pharmacol Ther. 2019;193:31–49. https://doi.org/10.1016/j.pharmthera.2018.08.006.
95. Paez-Ribes M, Gonzalez-Gualda E, Doherty GJ, Munoz-Espin D. Targeting senescent cells in translational medicine. EMBO Mol Med. 2019;11(12):e10234. https://doi.org/10.15252/emmm.201810234.
96. Krishnamurthy J, Torrice C, Ramsey MR, Kovalev GI, Al-Regaiey K, Su L, et al. Ink4a/Arf expression is a biomarker of aging. J Clin Invest. 2004;114(9):1299–307. https://doi.org/10.1172/JCI22475.
97. Wang E. Senescent human fibroblasts resist programmed cell death, and failure to suppress bcl2 is involved. Cancer Res. 1995;55(11):2284–92.
98. Zhu Y, Tchkonia T, Pirtskhalava T, Gower AC, Ding H, Giorgadze N, et al. The Achilles' heel of senescent cells: from transcriptome to senolytic drugs. Aging Cell. 2015;14(4):644–58. https://doi.org/10.1111/acel.12344.

99. Breccia M, Alimena G. Activity and safety of dasatinib as second-line treatment or in newly diagnosed chronic phase chronic myeloid leukemia patients. BioDrugs. 2011;25(3):147–57. https://doi.org/10.2165/11591840-000000000-00000.
100. Chan CM, Jing X, Pike LA, Zhou Q, Lim DJ, Sams SB, et al. Targeted inhibition of Src kinase with dasatinib blocks thyroid cancer growth and metastasis. Clin Cancer Res. 2012;18(13):3580–91. https://doi.org/10.1158/1078-0432.CCR-11-3359.
101. Zhu Y, Doornebal EJ, Pirtskhalava T, Giorgadze N, Wentworth M, Fuhrmann-Stroissnigg H, et al. New agents that target senescent cells: the flavone, fisetin, and the BCL-X(L) inhibitors, A1331852 and A1155463. Aging (Albany NY). 2017;9(3):955–63. https://doi.org/10.18632/aging.101202.
102. Kirkland JL, Tchkonia T. Senolytic drugs: from discovery to translation. J Intern Med. 2020;288(5):518–36. https://doi.org/10.1111/joim.13141.
103. Robbins PD, Jurk D, Khosla S, Kirkland JL, LeBrasseur NK, Miller JD, et al. Senolytic Drugs: Reducing Senescent Cell Viability to Extend Health Span. Annu Rev Pharmacol Toxicol. 2021;61:779–803. https://doi.org/10.1146/annurev-pharmtox-050120-105018.
104. Chang J, Wang Y, Shao L, Laberge RM, Demaria M, Campisi J, et al. Clearance of senescent cells by ABT263 rejuvenates aged hematopoietic stem cells in mice. Nat Med. 2016;22(1):78–83. https://doi.org/10.1038/nm.4010.
105. Zhu Y, Tchkonia T, Fuhrmann-Stroissnigg H, Dai HM, Ling YY, Stout MB, et al. Identification of a novel senolytic agent, navitoclax, targeting the Bcl-2 family of anti-apoptotic factors. Aging Cell. 2016;15(3):428–35. https://doi.org/10.1111/acel.12445.
106. Yousefzadeh MJ, Zhu Y, McGowan SJ, Angelini L, Fuhrmann-Stroissnigg H, Xu M, et al. Fisetin is a senotherapeutic that extends health and lifespan. EBioMedicine. 2018;36:18–28. https://doi.org/10.1016/j.ebiom.2018.09.015.
107. Arpon A, Riezu-Boj JI, Milagro FI, Marti A, Razquin C, Martinez-Gonzalez MA, et al. Adherence to Mediterranean diet is associated with methylation changes in inflammation-related genes in peripheral blood cells. J Physiol Biochem. 2016;73(3):445–55. https://doi.org/10.1007/s13105-017-0552-6.
108. Barnett JB, Dao MC, Hamer DH, Kandel R, Brandeis G, Wu D, et al. Effect of zinc supplementation on serum zinc concentration and T cell proliferation in nursing home elderly: a randomized, double-blind, placebo-controlled trial. Am J Clin Nutr. 2016;103(3):942–51. https://doi.org/10.3945/ajcn.115.115188.
109. Casas R, Sacanella E, Urpi-Sarda M, Chiva-Blanch G, Ros E, Martinez-Gonzalez MA, et al. The effects of the Mediterranean diet on biomarkers of vascular wall inflammation and plaque vulnerability in subjects with high risk for cardiovascular disease. A randomized trial. PLoS One. 2014;9(6):e100084. https://doi.org/10.1371/journal.pone.0100084.
110. Choi IY, Lee C, Longo VD. Nutrition and fasting mimicking diets in the prevention and treatment of autoimmune diseases and immunosenescence. Mol Cell Endocrinol. 2017;455:4–12. https://doi.org/10.1016/j.mce.2017.01.042.
111. De la Fuente M, Hernanz A, Guayerbas N, Victor VM, Arnalich F. Vitamin E ingestion improves several immune functions in elderly men and women. Free Radic Res. 2008;42(3):272–80. https://doi.org/10.1080/10715760801898838.
112. Meydani M, Macauley JB, Blumberg JB. Influence of dietary vitamin E, selenium and age on regional distribution of alpha-tocopherol in the rat brain. Lipids. 1986;21(12):786–91. https://doi.org/10.1007/BF02535413.
113. Lee J, Kim YS, Kim E, Kim Y, Kim Y. Curcumin and hesperetin attenuate D-galactose-induced brain senescence in vitro and in vivo. Nutr Res Pract. 2020;14(5):438–52. https://doi.org/10.4162/nrp.2020.14.5.438.
114. Sharifi-Rad J, Rayess YE, Rizk AA, Sadaka C, Zgheib R, Zam W, et al. Turmeric and Its Major Compound Curcumin on Health: Bioactive Effects and Safety Profiles for Food, Pharmaceutical, Biotechnological and Medicinal Applications. Front Pharmacol. 2020;11:01021. https://doi.org/10.3389/fphar.2020.01021.
115. Gupta SC, Patchva S, Aggarwal BB. Therapeutic roles of curcumin: lessons learned from clinical trials. AAPS J. 2013;15(1):195–218. https://doi.org/10.1208/s12248-012-9432-8.

116. McCubrey JA, Lertpiriyapong K, Steelman LS, Abrams SL, Yang LV, Murata RM, et al. Effects of resveratrol, curcumin, berberine and other nutraceuticals on aging, cancer development, cancer stem cells and microRNAs. Aging (Albany NY). 2017;9(6):1477–536. https://doi.org/10.18632/aging.101250.
117. Ramirez-Garza SL, Laveriano-Santos EP, Marhuenda-Munoz M, Storniolo CE, Tresserra-Rimbau A, Vallverdu-Queralt A, et al. Health Effects of Resveratrol: Results from Human Intervention Trials. Nutrients. 2018;10(12). https://doi.org/10.3390/nu10121892.

Chapter 10
Cellular Senescence

Tamas Fülöp, Alan A. Cohen, A. Larbi, and J. M. Witkowski

Introduction

In the 1960s, when Hayflick carried out his first experiments in cell cultures culminating in his theory of replicative senescence [1], nobody could have imagined that cellular senescence would become one of the most studied aspects of aging biology [2–5]. The Hayflick limit stated that cells in culture attain a limit beyond which they will not be able to further proliferate, yet they will not die [6]. Cells extracted from older adults performed fewer cycles than those from younger adults. This led at this time to the programmed cell aging theory. Later, it was discovered that this senescence was related to telomere attrition, as the telomeres shorten with each cell division [7]. Shortened telomeres are no longer able to protect the genetic material, and the cell as a protection enters a senescent state.

Later, this concept was further extended after numerous studies to the notion of the senescence associated secretory phenotype (SASP). This means that these cells are in cell cycle arrests, resist apoptosis but are still metabolically very active by secreting proinflammatory cytokines and other biologically active molecules [8–10]. Thus,

T. Fülöp (✉) · A. Larbi
Geriatric Division, Department of Medicine, Faculty of Medicine and Health Sciences, Research Center on Aging, University of Sherbrooke, Sherbrooke, QC, Canada
e-mail: Tamas.Fulop@Usherbrooke.ca

A. A. Cohen
Groupe de Recherche PRIMUS, Department of Family Medicine, University of Sherbrooke, 3001 12E Avenue N, Sherbrooke, QC J1H 5N4, Canada

Department of Environmental Health Sciences, Butler Columbia Aging Center, Columbia University, New York, NY, USA

A. Larbi
Beckman Coulter, Lyon, France

J. M. Witkowski
Department of Pathophysiology, Medical University of Gdansk, Gdansk, Poland

V. Bueno (ed.), *Cellular and Molecular Aspects of Ageing*,
https://doi.org/10.1007/978-3-031-55022-5_10

cellular senescence became considered as a major mechanism of aging, generating many investigations and even much hype for drug (called senolytic/xenomorphic) development or repurposing to eliminate senescent cells [11, 12]. Therefore, it is very important to understand the phenomenon of cellular senescence, including its process and its modulation in aging. This review will summarize our present knowledge on senescence and its role in immunosenescence, in inflammaging, aging and in cancer.

What Is Cellular Senescence and How Does It Occur?

There are several forms of cellular senescence defined basically as an irreversible cell cycle arrest to prevent proliferation of damaged cells [13]. One way to arrive at this cell cycle arrest is what was described by Hayflick as replicative senescence due to telomere shortening [1]; however, this can also be achieved by chronic infectious or antigenic stimulation like in case of immune cells. Other ways to arrive at a senescent cell state are related to intracellular or extracellular factors inducing nuclear or mitochondrial DNA (mtDNA) damage [14]. This persistent DNA damage response induces senescence in normal cells and represents a form of premature senescence [15]. Senescence has various initiators such as oncogenes, oncogene-induced senescence, irradiation, or oxidative stress [16, 17]. Nowadays, there are also therapy-related causes, such as senescence induced by chemotherapy or ionizing radiation [18, 19]. Among these chemotherapeutic agents for example the CDK4/6 inhibitors induced senescence-like cell cycle arrest demonstrated first in cell lines and finally implemented in clinics for oestrogen receptor-positive breast cancer treatment [20]. Therefore, cell senescence may occur due to various stressors occurring throughout life, either naturally or induced by therapy.

Most of these cells do not have very special or universal senescence markers, which is real challenge. However, most of the described stressor-induced senescent cells will express inhibitors of cyclin dependent kinases such as $p16^{INK4A}$ and $p21^{WAF1/CIP1}$ which respectively inhibit different cyclins which in turn will block the retinoblastoma (Rb) protein [21]. This latter will suppress the expression of E2F target genes resulting in senescent cycle arrest [22]. Some other senescence markers may be also used such as the p19ARF, p53, PAI-1 and p27 to characterize these cells [23]. However, the most used markers remain the p21 and p16.

Morphologically these senescent cells are flattened and present vacuolization [24] therefore very different from normal cells. If the senescence is induced by persistent DNA damage the cells contain DNA damage foci such as accumulation of 53BP1, phosphorylation of histone H2AX, or other DNA repair markers [25]. They may also exhibit senescence-associated heterochromatin foci formation, loss of lamin B1 upregulation of some miRNAsi [26]. Many others have been described but the ones mentioned are the most often found in senescent cells. One other common marker of the senescent cells is the increased activity of senescence associated β-galactosidase (SA-β-gal) [27]. All of these markers are specific for senescent cells but are not necessarily present in all senescent cells. Therefore, several marker combinations should be used to identify senescent cells. With the progress of research new and more specific markers are constantly discovered.

However, what is emerging from recent studies is that the alteration of cellular metabolism is also essential for arriving at the accomplishment of the whole senescent programme of the cells. One of the characteristics of this alteration is the depletion of the catabolic enzyme glycogen phosphorylase leading to glycogen accumulation and concomitantly to increased glucose consumption and lactate production and ultimately involving the mitochondria [28, 29].

Multiple Physiological Roles of Senescence

Physiologically senescence is an anti-cancer process eliminating damaged cells which could become tumorigenic were the DNA damage to persist [30]. However, in many cases the development of senescent cells is the consequence of DNA alterations, and the cells will survive and even become very active by secreting various pro-inflammatory mediators, exhibiting the senescence associated secretory phenotype (SASP). Furthermore, cellular senescence plays a beneficial role in tissue remodeling during embryogenesis. The development and function of these cells occurs very differently from one organ to the other, leading to coordinated morphogenesis/organogenesis [31].

Cellular senescence may also play a favorable role in tissue repair [32, 33]. This consists of four very well-orchestrated primary phases: hemostasis, inflammation, proliferation, and remodeling [34]. Cellular senescence by the emergence of SASP may influence each phase of tissue repair/wound healing. These factors may drive the differentiation of fibroblasts into myofibroblasts to contract the wound [35]. Related to wound healing, cellular senescence may influence fibrosis either by limiting it as in the healing process of myocardial infarction or by accelerating it as in idiopathic pulmonary fibrosis [36]. Cellular reprogramming may be an important process in tissue regeneration; however, it is not fully understood whether and how cellular senescence influences this process either favorably by favoring iPS cells or detrimentally by inhibiting the regeneration [37]. Experimentally, NANOG and LIN28 may counteract the senescence programme especially during cellular reprogramming [38]. The LIN28 is able to bypass the senescence programme by translation of HMGA2, a transcriptional repressor of INK4A/ARF [39]. Together this indicates that cellular senescence effects may always depend on the environment where they occur.

Cellular senescence is generally considered as irreversible cell cycle arrest; however, recently some studies demonstrated that a small population of senescent cells may override this senescence and may re-enter the cell cycle under special conditions [40]. This could be a very important phenomenon, as have been seen for exhausted immune cells under immunotherapy, suggesting that they may be target for therapeutic rescue in various pathologic situations such as cancer.

Molecular Mechanisms of Senescence and SASP

The molecular mechanisms underlying telomere shortening-induced senescence involve the DNA damage response (DDR), which includes the activation of ataxia-telangiectasia mutated (ATM) or ATM—and Rad-3 related (ATR) kinases [41]. These pathways will block the cell cycle progression by stabilizing p53 and the Cdk inhibitor p21. This pathway may at the beginning create a transient proliferation arrest which will help the cellular machinery to repair the alteration, but in cases when this is not successful the cell may evolve towards senescence for not propagating the damage. Beside telomere shortening, other physiological stresses such as increased oxidative stress, endoplasmic reticulum stress, IFN related responses, irradiation, anticancer chemotherapeutic agents, oncogenes and the immune response may lead to senescence in normal or cancer cells [42, 43]. A systemic inflammatory state, mainly chronic ones, may also lead to cellular senescence. While there are many well-established causes of senescence, from the molecular and morphological point of view the distinction between the various types of senescence is challenging.

What are the intracellular molecular mechanisms in these senescent cells for sustained pro-inflammatory mediator production? The senescent cells are secreting a plethora of molecules playing a role in the modulation of the immune response, having pro-inflammatory properties, growth factors, chemokines and proteases, commonly called the SASP [44]. Senescence may be initiated by these mediators in a paracrine or autocrine manner [45]. Extracellularly, IL-1β may increase cellular senescence with the acquisition of the SASP in a paracrine manner [46]. The NAD and the mTOR pathways also regulate the SASP through nutrient sensing. IFN is also able to trigger senescence by inducing DNA damage in the target cells. Among the intracellular hubs, the NFkB transcription factor as well as the CCAAT/enhancer binding protein IC (C-EBP-β) seem to play an important role [47]. There are a myriad of mediators which may further induce senescence. These findings raise the question of why natural selection has produced so many factors which can induce senescence. One explanation may be that senescence from an evolutionary point of view is so important, whatever the age of the subject, that it could not be left to chance to determine whether damaged cells will be eliminated, even at the expense of development of negative consequences of SASP. For this reason, redundant [48] or degenerate [49] mechanisms have evolved to ensure senescence is triggered when the conditions are appropriate. Even SASP per se may be very important for immune response regulation and for the control of the tissue microenvironment. This knowledge led to the tentative therapeutic reprogramming of SASP in the cancer environment to boost the immune response or in contrast in an aging environment to eliminate or inhibit SASP [50, 51].

It has also been shown that the stimulation of main pattern recognition receptors (PRR) the TLRs can also result in SASP induction [52]. Intracellular TLRs act as damage-associated molecular pattern recognition receptors, especially for intracellularly derived DNA but also for intracellular viral or bacterial DNA products. Regardless of the stimulating agent, it results in the NFkB stimulation resulting in

SASP. This process may be an important part of innate host defense highlighting the situation dependent role and consequences of SASP in the organism. These stimulations lead to the epigenetic modulation of expression of the genes of the SASP factors [53]. DNA damage modulates the proteasomal degradation of one of the major histones H3K79 dimethyltransferase as well as others, resulting in increased SASP factor gene expression [54].

Recently another pathway, the cGAS-STING pathway, has been shown to be involved in cellular senescence and SASP [55]. This is usually considered an antiviral pathway that results in the IFN gene expression with ultimate IFN type I and III production [56]. The cGAS is sensing the intracellular presence of pathogenic e.g. viral DNA. Once this recognition occurs it will generate the cyclic GMP-AMP complex (cGAMP). The cGAMP will stimulate as second messenger STING which further recruits after autophosphorylation the TANK binding kinase 1 (TBK1). TBK1 will phosphorylate the interferon regulatory, factor 3 (IFR3) which further stimulates the transcription of type-I interferon (IFN). Interestingly TBK1 also activates NFKB leading to the expression of IFN and the other pro-inflammatory medicators [57, 58]. This pathway, a part of the innate immune system to fight infections, became an important mediator of cellular senescence [59]. cGAS is also able to detect altered or generated self-DNA which will initiate the same cascade as viral DNA, leading to SASP in senescent cells [60]. These DNA products are numerous and constantly generated and of course do not lead to senescence, but the circumstances in the microenvironment likely reinforce the deleterious effects of these DNA products. Among these products there are cytoplasmic chromatin fragments (CCF), mtDNA, cDNA, micronuclei and nuclear buds [61]. Physiologically, these fragments are destroyed by DNases. Interestingly, the downregulation of DNAses which are transcriptional target of the E2F complex results in the accumulation of these DNA products [62]. This reinforces the already altered signaling in the senescent cells as mentioned above. There is another potential mechanism to increase the signaling through cGAMP pathway. This concerns the release of exosomes from cancer or senescent cells via the SLC19A1, resulting in the activation of STING in the recipient cells [63]. The role of the cGAS-STING initiated pathway is gaining more and more importance as this is a druggable pathway and already both agonists and antagonists are developed and are in clinical trials e.g. in cancer.

Senescent Cells and Aging

One of the most modern and current theories of aging states that aging is the consequence of the accumulation of senescent cells even in normal and physiological circumstances, causing progressive functional decline [64]. Senescent cells also accumulate at sites of age-related pathologies, especially in arteries (atherosclerosis) and the joints (osteoarthritis) as especially studied but also in many other tissues [65, 66]. As mentioned, the senescence programme has multiple physiological roles, perhaps most importantly to prevent the occurrence of cancers due to DNA alterations of any kind [67]. As a side effect, it seems, based mainly on animal experiments, that during life there is an increasing number of senescent cells in various organs jamming the older tissues and propagating the SASP. This condition could be representative of one of the most popular evolutionary theories of aging, the antagonistic pleiotropy theory [68, 69]. This states that natural selection does not eliminate what is beneficial at younger ages even if it becomes detrimental in older age because the benefit (the reproductive success) is higher that the damage. There is no real selection pressure to eliminate this process during aging. However, it is an open question whether senescence is purely detrimental with aging or whether it also represents an adaptation through aging to decrease the emergence of cancer as much as possible until the end of human life. This is plausible; as was demonstrated by the Franceschi group, cancer incidence sharply decreases after 85 years of age [70]. Furthermore, we do not know whether the contribution of SASP per se is detrimental or may sustain some threshold of activity as it is the case in centenarians and semi-supercentenarians [71, 72]. Finally, what is true in mice may not be true in humans, or what is true in lab mice may not be true in all mice, and perhaps that is why attempts to decrease senescent cells in humans have not yet shown clear health benefits.

Another argument for the detrimental aging effect of senescent cells and SASP is their putative role in age-dependent diseases. From the geroscience point of view, aging and age-related diseases are caused by the same molecular mechanism, so treating aging may avoid or retard the occurrence of age-related diseases [73]. From the cellular senescence angle, which is now part of the hallmarks of aging, these cells should be everywhere in aging tissues contributing to diseases [74]. It is of note that most of the senescence-associated knowledge and conclusions are derived from mice studies. Indeed, the successful in vivo elimination of the senescent cells from mice leads to the increase of their lifespan and healthspan [75]. In humans, most of the assumption that these cells exist come form some studies describing the presence of cellular senescence markers in tissues or in cell cultures without really translating these markers into functions.

It is believed that senescent cells play a major role in atherosclerosis as VSMCs and endothelial cells showed upregulation of cellular senescence markers in human atherosclerotic plaques [76]. However, several studies showed that certain genetic associations with cellular senescence may be protective instead of being detrimental. The senescence of monocytes and macrophages may limit the growth of the plaques, but SASP may participate in disease progression and plaque destabilisation [77, 78]. Therefore, it is difficult to conclude that in human atherosclerosis is related to the increased number of senescent cells in the intima and media, however this is certainly related to the chronic arterial wall inflammation [79].

One other human disease, which is not entirely an age-related disease as this may occur very early in age even in middle aged adults [80], considered to be associated with the accumulation of senescent cells is osteoarthritis. This assumes that chondrocytes related to the disease show typical senescence markers. Characteristically these data originate mainly from mice and the elimination of these cells improved osteoarthritis. The destruction of the cartilage is related to SASP factors such as MMPs, oxidative stress and pro-inflammatory cytokines. However, all these factors may originate from a physiological chronic inflammatory process, as in the case of atherosclerosis.

Recently osteoporosis has been also related to the existence of senescent cells [81]. Senescent bone cells increase osteoclast progenitor survival while inhibiting osteoblast activity. This results in bone resorption. However, many other hormonal and environmental stress factors may induce osteoporosis.

Neurodegeneration has also been related to the existence of senescent cells based on the argument that as the main risk factor is aging and age is related to cellular senescence therefore the senescent cells should also play an important role in neurodegeneration [82]. In humans presently there exists very little experimental evidence whether a causal link exists between cellular senescence and neurodegeneration. The link is in most cases speculative and a sort of circular and reductionist conceptualization of the neurodegenerative diseases. The most important speculative argument is that indeed neuroinflammation plays an important role in neurodegeneration and as SASP is nowadays considered as the driving molecular foundation of inflammaging, cellular senescence should play a role via SASP [83]. The main cells in the brain showing markers of senescence are astrocytes either in AD or PD [84, 85]. In mice, elimination of senescent cells from the brain was sufficient to improve SAD and PD.

Together presently there are no solid experimental data in humans to show that senescent cells really accumulate with aging, although this seems likely based on the nature of senescent cells. Even more the relationship in humans especially in vivo between the so-called age-related diseases and senescent cells is more than questionable, as most of these conclusions came from animal and in vitro experiments. Because of the complexity between the origin, the pathways, the roles, and the microenvironment, it is very difficult to assign a clear pathological role in humans to cellular senescence. More experimental evidence is needed in human despite all of the overwhelming evidence in mice.

Senescence in the Immune System

It is widely accepted that with aging the immune system changes [86–88]. Most of the time it is considered as a deleterious process. Mainly it is related to, as the word immunosenescence reflects, the development of senescence over time in immune cells. Therefore, the immune system is considered as the representative prototype of the cellular senescence process during aging.

In the innate immune system, the main suspects are macrophages; however, the presence of senescence in them is still a debate [89]. There is evidence mostly in animal models that microglia may enter into senescence during aging. With aging these cells and also macrophages seem to shift towards more anti-inflammatory phenotypes which may be considered as senescent. The subpopulation shift in monocytes was also related to monocyte senescence mainly in the non-classical subpopulation [90]. However recent data indicate that the expression of p16 in myeloid cells instead of generating senescence may be necessary for their differentiation and polarization, which could be in line with the adaptive concept of the innate immune response with aging [91].

In the adaptive immune system, a huge literature exists to describe, define, and characterize the function of senescent T cells due to their continuous stimulation and proliferation during life [92–94]. These T cells may become either exhausted or senescent [95]. There exist many commonalities between these two cell types, but the exhausted cells are not metabolically active, in contrast to senescent cells. The exhausted T cells can be reactivated to be highly functional by checkpoint inhibition, while senescent cells after our present knowledge do not restore immunological function even if they are able to secrete and express high levels of senescence related immunological markers.

Senescent T lymphocytes, besides expressing the traditional senescence markers, possess a characteristic secretome including IL6, IL8, TGF-β, IFN-γ as well as the downregulation of important co-stimulatory surface markers including the most important, CD28 [96, 97]. The lifelong proliferation of these cells not only induce senescence, but also the inflammatory milieu related to aging or cancer. In this inflammatory milieu TNF and IFN may induce senescence by activating the p38MAPK family member and downregulating telomerase concomitantly to the decrease of CD28 expression [98]. This leads to the accumulation of CD8+CD28- T cells with aging and in age-related diseases, often considered as SASP and participating in pathogenesis and the decrease of the immune response to infections and vaccination in older subjects [99]. However, new data suggest that the simultaneous expression on the surface of these senescent or CD8+CD28- T cells the NK cells characteristics, these cells have a very important immune positive functional compensatory role mainly with advancing age, such as to reduce the cancer occurrence [100]. Moreover, these CD28- T cells in the CD4 T cells may skew them into the regulatory phenotype and function [101]. These Tregs may in turn either induce senescence as part of the immune surveillance or induce senescence in responder T cells.

The most recent contribution of immune senescence is the implication of SASP as a main driver of inflammaging [102]. This is one other way to implicate the senescence of immune cells into propagation of senescence and the stimulation of the adaptive immune compartment. However, the distinction between SASP and inflammaging is not yet clear. At one extreme, perhaps inflammaging is simply the accumulation of SASP—there is a fair amount of overlap in the markers of the two, and no investigation has systematically analyzed them all in humans. On the other, perhaps they are two quite distinct processes with several overlapping markers. In the middle, and perhaps most likely, the SASP may be one factor contributing to inflammaging, but not the only one, and there may be feedback loops in which SASP triggers more generalized inflammaging, and inflammaging triggers some cellular senescence and thereby SASP.

Together, at the beginning the immune system can eliminate senescent cells, but later as this is a propagating process the immune system will be subverted and will even contribute to the development of other senescent cells. However, the extent and the role of these processes is probably highly variable depending on the internal and environmental factors, making very difficult the assumption on their respective role in the general physiology or specific pathologies of the organism.

Senescent Cells and Cancer

As mentioned, senescence is a very powerful anti-cancer process normally occurring continuously in the organism such as in pre-malignant tumors or after a senescence-inducing stressor such as oxidative stress. It should be also mentioned, which will be treated below, that senescent cells accumulate as the cancer pressure increases with age (time), as was shown for aged mice.

One of the most common forms of senescence in the context of cancer is called oncogene-induced senescence (OIS) [103, 104]. This process is found in many human cancers and mediated via mutations in the RAS oncogene [105]. It was shown that this sole mutation is not sufficient to induce tumorigenic transformation because in contrast it results in growth arrest as the cells concomitantly express p19ARF, Pml, p53, Rb and p16 [106]. Other mutations and activations are necessary to override this RAS mutation. The most important components may be c-MYC, E1A or DRIL1 [107]. Beside RAS other oncogenes such as HER2, EGFR or BRAF may also induce senescence [108]. Together the oncogene-induced senescence is a complex cell autonomous regulation of cellular senescence requiring the cooperation of many different intracellular factors. Inactivation of certain oncogene may also lead to senescence, such as the inactivation of MYC [109]. This requires the concerted action of several mechanisms in the tumor microenvironment such as the activation of the immune system. The loss of PTEN also induces cell senescence by the expression of senescence markers such as p16 [110]. Together, cellular senescence is a very complex and highly regulated process either to promote or to avoid cancer.

The discovery of the SASP changed the biological landscape of cancer, aging, and age-related diseases, being one of the major fuels for inflammation or inflammaging. These cells secrete pro-inflammatory cytokines, chemokines, growth factors and MMPs [111]. SASP is a double aged sword in the cancer microenvironment as they can act on neighbouring cells and on the recruitment and activation of immune cells resulting in both antitumorigenic and tumor promoting effects [112]. Therefore, these mediators may have several physiological roles. In the context of cancer, these cells are able to induce senescence in other cells in a paracrine and autocrine manner, thereby exerting a more extended anti-tumor effect [113]. This will result in the decrease in the active tumor cell population able to grow and metastasize.

One of the other essential roles of senescent cells is to recruit immune cells from the innate and adaptive arms, stimulating them to clear the senescent and proliferating tumor cells [30]. This is termed senescence-surveillance [114]. This a very powerful way to eliminate senescent tumor cells as well as concomitantly oncogene-expressing cells. Probably at the very beginning of the tumor emergence this SASP process play a crucial role in the immune-senescence-surveillance maintaining the tumors at a clinically silent state. Indeed, the Th1 lymphocytes may promote senescence in tumor cells by secreting IFN-γ and TNF-α [115]. The macrophages secreting TGF-β may also induce senescence in tumor cells [116]. This activation of the immune surveillance by SASP may be a powerful anticancer process.

Despite these beneficial effects, senescent cells may become detrimental in the long run by their accumulation and constant secretion of SASP related substances. This is the process involved for development of age-related diseases such as atherosclerosis and cancer. This has several causes. One is the decrease in immune system function resulting in decreased senescence surveillance leading to the accumulation of senescent cells [117]. The main SASP secreted substances such as IL-1β or MMP have been demonstrated to contribute to cancer development by increasing proliferation and driving tumor vascularization [118]. These senescent cells may also accumulate after chemotherapy and may favor tumor progression with time [119]. The immunosuppressive immune cells such myeloid derived suppressor cells (MDSC) may also be recruited by SASP [120]. The secretion of IL-8 by senescent fibroblasts in the tumor microenvironment promotes pancreatic cancer cell invasion and metastasis [121]. There are many other mediators released either by tumor senescent cells or senescent cells in the tumor microenvironment presenting SASP contributing to tumor growth and dissemination.

The above described cGAS-STING pathway also became important in cancer [55–60]. For several tumors the chronic inflammation induced by SASP via the cGAS-STING pathway seems to be essential for their development such as HCC, lung cancer [62, 122]. The inverse is also true that the inhibition of cGAS-STING pathway impaired the immune-surveillance of senescent cells and therefore contributed to tumorigenesis. The biological effect of cGAS-STING thus depends on the biological context [123]. In acute circumstances, the SASP factors may be beneficial but on a chronic basis they will induce various diseases including cancer.

Senescent Cell Clearance

The main actors of the elimination of senescent cells are the innate immune cells, M1 macrophages, NK cells and the adaptive immune cells, Th1 lymphocytes [124]. NK cells recognize the senescent cells by the expression of specific surface receptors such as NKG2D, MICA or ULBP2. They are killed by granzyme or perforin production leading to cytotoxicity [125]. The expression of p53 in senescent cells may drive the polarization of macrophages towards M1.

Senescent cells may also be eliminated by immunotherapy. Immunotherapy may consist of vaccination, T cell reactivation or autologous infusion of T cells such as CD4+ Th1 cells or CAR-T cells therapy which eventually may induce senescence in cancer [126]. Most of these therapies are still experimental but in clinic some trials are performed. The most promising are presently the CAR-T and CAR-NK cell therapies.

Potential Interventions

Cancer

Many drugs already in clinical use in cancer treatment such as docetaxel, vincristine, etoposide, or cisplatin will target and induce the therapy-induced senescence [18]. Radiotherapy may also induce cell senescence in tumor cells by DNA damage enhancement and may contribute to what is called the abscopal effect helping the elimination of these senescent cells by the immune system [127]. The CDK4/6 inhibitor palbociclib is very efficiently used in clinic as described above.

There is substantial active research to find modulators of cellular senescence either in the context of cancer, aging, and age-related diseases. In cancer, there are many different tracks to induce senescence such as telomerase inhibition, modulation of the cell-cycle machinery, the modulation of p53 and MYC, SIRT1 inhibitors, immunotherapy such as for inhibition of MDSCs recruitment and function and SASP reprogramming [128].

Recent research in the domain of SASP reprogramming reinforcing the protective and anti-tumorigenic role of SASP gave promising results. JAK2 inhibitors, already used in clinic, may induce SASP reprogramming leading to the reactivation of the senescence immune surveillance [129]. The inhibition of the mTOR pathway, e.g. by rapamycin, will prevent the pro-tumorigenic effects of SASP [130].

Senolytics and Senomorphics

Recent research aims to find means to selectively eliminate senescent cells to avoid the deleterious effects of the treatment-induced senescent cells and their SASP. The resistance of the senescent cells to apoptosis is derived from the over-expression of Bcl-2 family proteins; therefore, inhibition of Bcl-2 may induce apoptosis in these senescent cells [131]. There are many promising drugs; one already in human use is navitoclax, but unfortunately it produces many side effects which restrict its use. The combination of Dasatinib and Quercetin has been showing promising effects in vitro mainly in age-related diseases in mice, so far without being translated to humans [132]. There are many other substances in the pipeline based on the molecular knowledge of the senescence markers and SASP such as FOXO4 inhibitors, HSP90 inhibitors, and NAD+ precursors [133]. The field of senolytic/senomorphic drugs discovery is very active and may become when most of the molecular mechanisms and functional role of senescent cells are better understand in human setting.

Conclusion

Cellular senescence is a very complex process having physiological and pathological functions. Recently many data have been generated to better understand this biological phenomenon. This led the unraveling of the markers and as the functional hall marks of senescence. Nevertheless, there is an urgent need to better understand cellular senescence in the context of human biology.

It is now well established that SASP factors depending on the biological milieu and the secretion timeline may be tumor suppressor or tumorigenic. The development of senolytic drugs may be efficient to eliminate SASP cells but, as their long-term effects are not known, they are not likely to be clinically useful in the near future. Identical considerations may be applied to aging where their role is concretely unknown and therefore the interventions are also most hazardous.

The senescence phenomenon is a very important biological process when appropriately understood not only in a reductionist way, but also from a complex systems perspective, implying it may be either inhibited or enhanced to obtain the most valuable effects for health and function span of humans.

References

1. Hayflick L, Moorhead PS. The serial cultivation of human diploid cell strains. Exp Cell Res. 1961 Dec;25:585-621. https://doi.org/10.1016/0014-4827(61)90192-6.
2. von Zglinicki T. Role of oxidative stress in telomere length regulation and replicative senescence. Ann N Y Acad Sci. 2000 Jun;908:99-110. https://doi.org/10.1111/j.1749-6632.2000.tb06639.x.
3. Calcinotto A, Kohli J, Zagato E, Pellegrini L, Demaria M, Alimonti A.Cellular Senescence: Aging, Cancer, and Injury. Physiol Rev. 2019 Apr 1;99(2):1047-1078. https://doi.org/10.1152/physrev.00020.2018.
4. Campisi J. Aging, cellular senescence, and cancer. Annu Rev Physiol. 2013;75:685-705. https://doi.org/10.1146/annurev-physiol-030212-183653.
5. Hernandez-Segura A, Nehme J, Demaria M. Hallmarks of Cellular Senescence. Trends Cell Biol. 2018 Jun;28(6):436-453. https://doi.org/10.1016/j.tcb.2018.02.001.
6. Shay JW, Wright WE. Hayflick, his limit, and cellular ageing. Nat Rev Mol Cell Biol. 2000 Oct;1(1):72-6. https://doi.org/10.1038/35036093.
7. Kalmykova A. Telomere Checkpoint in Development and Aging. Int J Mol Sci. 2023 Nov 5;24(21):15979. https://doi.org/10.3390/ijms242115979.
8. Lopes-Paciencia S, Saint-Germain E, Rowell MC, Ruiz AF, Kalegari P, Ferbeyre G. The senescence-associated secretory phenotype and its regulation. Cytokine. 2019 May;117:15-22. https://doi.org/10.1016/j.cyto.2019.01.013.
9. Giroud J, Bouriez I, Paulus H, Pourtier A, Debacq-Chainiaux F, Pluquet O. Exploring the Communication of the SASP: Dynamic, Interactive, and Adaptive Effects on the Microenvironment. Int J Mol Sci. 2023 Jun 28;24(13):10788. https://doi.org/10.3390/ijms241310788.
10. Roger L, Tomas F, Gire V. Mechanisms and Regulation of Cellular Senescence. Int J Mol Sci. 2021 Dec 6;22(23):13173. https://doi.org/10.3390/ijms222313173.
11. Ohtani N. The roles and mechanisms of senescence-associated secretory phenotype (SASP): can it be controlled by senolysis? Inflamm Regen. 2022 Apr 2;42(1):11. https://doi.org/10.1186/s41232-022-00197-8.

12. Cuollo L, Antonangeli F, Santoni A, Soriani A. The Senescence-Associated Secretory Phenotype (SASP) in the Challenging Future of Cancer Therapy and Age-Related Diseases. Biology (Basel). 2020 Dec 21;9(12):485. https://doi.org/10.3390/biology9120485.
13. Regulski MJ. Cellular Senescence: What, Why, and How. Wounds. 2017 Jun;29(6):168-174.
14. Di Micco R, Fumagalli M, Cicalese A, Piccinin S, Gasparini P, Luise C, Schurra C, Garre' M, Nuciforo PG, Bensimon A, Maestro R, Pelicci PG, d'Adda di Fagagna F. Oncogene-induced senescence is a DNA damage response triggered by DNA hyper-replication. Nature. 2006 Nov 30;444(7119):638–42. https://doi.org/10.1038/nature05327.
15. Campisi J, d'Adda di Fagagna F. Cellular senescence: when bad things happen to good cells. Nat Rev Mol Cell Biol. 2007 Sep;8(9):729–40. https://doi.org/10.1038/nrm2233.
16. Gorgoulis V, Adams PD, Alimonti A, Bennett DC, Bischof O, Bishop C, Campisi J, Collado M, Evangelou K, Ferbeyre G, Gil J, Hara E, Krizhanovsky V, Jurk D, Maier AB, Narita M, Niedernhofer L, Passos JF, Robbins PD, Schmitt CA, Sedivy J, Vougas K, von Zglinicki T, Zhou D, Serrano M, Demaria M. Cellular Senescence: Defining a Path Forward. Cell. 2019 Oct 31;179(4):813-827. https://doi.org/10.1016/j.cell.2019.10.005.
17. Nousis L, Kanavaros P, Barbouti A. Oxidative Stress-Induced Cellular Senescence: Is Labile Iron the Connecting Link? Antioxidants (Basel). 2023 Jun 10;12(6):1250. https://doi.org/10.3390/antiox12061250.
18. Prasanna PG, Citrin DE, Hildesheim J, Ahmed MM, Venkatachalam S, Riscuta G, Xi D, Zheng G, Deursen JV, Goronzy J, Kron SJ, Anscher MS, Sharpless NE, Campisi J, Brown SL, Niedernhofer LJ, O'Loghlen A, Georgakilas AG, Paris F, Gius D, Gewirtz DA, Schmitt CA, Abazeed ME, Kirkland JL, Richmond A, Romesser PB, Lowe SW, Gil J, Mendonca MS, Burma S, Zhou D, Coleman CN. Therapy-Induced Senescence: Opportunities to Improve Anticancer Therapy. J Natl Cancer Inst. 2021 Oct 1;113(10):1285-1298. https://doi.org/10.1093/jnci/djab064.
19. Lee S, Lee JS. Cellular senescence: a promising strategy for cancer therapy. BMB Rep. 2019 Jan;52(1):35-41. https://doi.org/10.5483/BMBRep.2019.52.1.294.
20. Braal CL, Jongbloed EM, Wilting SM, Mathijssen RHJ, Koolen SLW, Jager A. Inhibiting CDK4/6 in Breast Cancer with Palbociclib, Ribociclib, and Abemaciclib: Similarities and Differences. Drugs. 2021 Feb;81(3):317-331. https://doi.org/10.1007/s40265-020-01461-2.
21. Takeuchi S, Takahashi A, Motoi N, Yoshimoto S, Tajima T, Yamakoshi K, Hirao A, Yanagi S, Fukami K, Ishikawa Y, Sone S, Hara E, Ohtani N. Intrinsic cooperation between p16INK4a and p21Waf1/Cip1 in the onset of cellular senescence and tumor suppression in vivo. Cancer Res. 2010 Nov 15;70(22):9381-90. https://doi.org/10.1158/0008-5472.CAN-10-0801.
22. Engeland K. Cell cycle regulation: p53-p21-RB signaling. Cell Death Differ. 2022 May;29(5):946-960. https://doi.org/10.1038/s41418-022-00988-z.
23. Calcinotto A, Alimonti A. Aging tumour cells to cure cancer: "pro-senescence" therapy for cancer. Swiss Med Wkly. 2017 Jan 17;147:w14367.
24. Herranz N, Gil J. Mechanisms and functions of cellular senescence. J Clin Invest. 2018 Apr 2;128(4):1238-1246. https://doi.org/10.1172/JCI95148.
25. Bártová E, Legartová S, Dundr M, Suchánková J. A role of the 53BP1 protein in genome protection: structural and functional characteristics of 53BP1-dependent DNA repair. Aging (Albany NY). 2019 Apr 17;11(8):2488–2511. https://doi.org/10.18632/aging.101917.
26. Richter T, von Zglinicki T. A continuous correlation between oxidative stress and telomere shortening in fibroblasts. Exp Gerontol. 2007 Nov;42(11):1039-42. https://doi.org/10.1016/j.exger.2007.08.005.
27. Lee BY, Han JA, Im JS, Morrone A, Johung K, Goodwin EC, Kleijer WJ, DiMaio D, Hwang ES. Senescence-associated beta-galactosidase is lysosomal beta-galactosidase. Aging Cell. 2006 Apr;5(2):187-95. https://doi.org/10.1111/j.1474-9726.2006.00199.x.
28. Liu B, Meng Q, Gao X, Sun H, Xu Z, Wang Y, Zhou H. Lipid and glucose metabolism in senescence. Front Nutr. 2023 Aug 23;10:1157352. https://doi.org/10.3389/fnut.2023.1157352.
29. Dou X, Fu Q, Long Q, Liu S, Zou Y, Fu D, Xu Q, Jiang Z, Ren X, Zhang G, Wei X, Li Q, Campisi J, Zhao Y, Sun Y. PDK4-dependent hypercatabolism and lactate production of

senescent cells promotes cancer malignancy. Nat Metab. 2023 Nov;5(11):1887-1910. https://doi.org/10.1038/s42255-023-00912-w.

30. Kang TW, Yevsa T, Woller N, Hoenicke L, Wuestefeld T, Dauch D, Hohmeyer A, Gereke M, Rudalska R, Potapova A, Iken M, Vucur M, Weiss S, Heikenwalder M, Khan S, Gil J, Bruder D, Manns M, Schirmacher P, Tacke F, Ott M, Luedde T, Longerich T, Kubicka S, Zender L. Senescence surveillance of pre-malignant hepatocytes limits liver cancer development. Nature. 2011 Nov 9;479(7374):547-51. https://doi.org/10.1038/nature10599.
31. Ritschka B, Storer M, Mas A, Heinzmann F, Ortells MC, Morton JP, Sansom OJ, Zender L, Keyes WM. The senescence-associated secretory phenotype induces cellular plasticity and tissue regeneration. Genes Dev. 2017 Jan 15;31(2):172-183. https://doi.org/10.1101/gad.290635.
32. Huang W, Hickson LJ, Eirin A, Kirkland JL, Lerman LO. Cellular senescence: the good, the bad and the unknown. Nat Rev Nephrol. 2022 Oct;18(10):611-627. https://doi.org/10.1038/s41581-022-00601-z.
33. Antelo-Iglesias L, Picallos-Rabina P, Estévez-Souto V, Da Silva-Álvarez S, Collado M. The role of cellular senescence in tissue repair and regeneration. Mech Ageing Dev. 2021 Sep;198:111528. https://doi.org/10.1016/j.mad.2021.111528.
34. Flanagan M. The physiology of wound healing. J Wound Care. 2000 Jun;9(6):299–300. https://doi.org/10.12968/jowc.2000.9.6.25994.
35. Demaria M, Ohtani N, Youssef SA, Rodier F, Toussaint W, Mitchell JR, Laberge RM, Vijg J, Van Steeg H, Dollé ME, Hoeijmakers JH, de Bruin A, Hara E, Campisi J. An essential role for senescent cells in optimal wound healing through secretion of PDGF-AA. Dev Cell. 2014 Dec 22;31(6):722-33. https://doi.org/10.1016/j.devcel.2014.11.012.
36. Meyer K, Hodwin B, Ramanujam D, Engelhardt S, Sarikas A. Essential Role for Premature Senescence of Myofibroblasts in Myocardial Fibrosis. J Am Coll Cardiol. 2016 May 3;67(17):2018-28. https://doi.org/10.1016/j.jacc.2016.02.047.
37. Lapasset L, Milhavet O, Prieur A, Besnard E, Babled A, Aït-Hamou N, Leschik J, Pellestor F, Ramirez JM, De Vos J, Lehmann S, Lemaitre JM. Rejuvenating senescent and centenarian human cells by reprogramming through the pluripotent state. Genes Dev. 2011 Nov 1;25(21):2248-53. https://doi.org/10.1101/gad.173922.111.
38. Hanna J, Saha K, Pando B, van Zon J, Lengner CJ, Creyghton MP, van Oudenaarden A, Jaenisch R. Direct cell reprogramming is a stochastic process amenable to acceleration. Nature. 2009 Dec 3;462(7273):595-601. https://doi.org/10.1038/nature08592.
39. Xu B, Zhang K, Huang Y. Lin28 modulates cell growth and associates with a subset of cell cycle regulator mRNAs in mouse embryonic stem cells. RNA. 2009 Mar;15(3):357-61. https://doi.org/10.1261/rna.1368009.
40. Coutelier H, Xu Z. Adaptation in replicative senescence: a risky business. Curr Genet. 2019 Jun;65(3):711-716. https://doi.org/10.1007/s00294-019-00933-7.
41. d'Adda di Fagagna F, Reaper PM, Clay-Farrace L, Fiegler H, Carr P, Von Zglinicki T, Saretzki G, Carter NP, Jackson SP. A DNA damage checkpoint response in telomere-initiated senescence. Nature. 2003 Nov 13;426(6963):194–8. https://doi.org/10.1038/nature02118.
42. Chen Q, Fischer A, Reagan JD, Yan LJ, Ames BN. Oxidative DNA damage and senescence of human diploid fibroblast cells. Proc Natl Acad Sci U S A. 1995 May 9;92(10):4337-41. https://doi.org/10.1073/pnas.92.10.4337.
43. Packer L, Fuehr K. Low oxygen concentration extends the lifespan of cultured human diploid cells. Nature. 1977 Jun 2;267(5610):423-5. https://doi.org/10.1038/267423a0.
44. Rodier F. Detection of the senescence-associated secretory phenotype (SASP). Methods Mol Biol. 2013;965:165-73. https://doi.org/10.1007/978-1-62703-239-1_10.
45. Parikh P, Wicher S, Khandalavala K, Pabelick CM, Britt RD Jr, Prakash YS. Cellular senescence in the lung across the age spectrum. Am J Physiol Lung Cell Mol Physiol. 2019 May 1;316(5):L826-L842. https://doi.org/10.1152/ajplung.00424.2018.
46. Hubackova S, Krejcikova K, Bartek J, Hodny Z. IL1- and TGFβ-Nox4 signaling, oxidative stress and DNA damage response are shared features of replicative, oncogene-induced, and drug-induced paracrine 'bystander senescence'. Aging (Albany NY). 2012 Dec;4(12):932–51. https://doi.org/10.18632/aging.100520.

47. Chien Y, Scuoppo C, Wang X, Fang X, Balgley B, Bolden JE, Premsrirut P, Luo W, Chicas A, Lee CS, Kogan SC, Lowe SW. Control of the senescence-associated secretory phenotype by NF-κB promotes senescence and enhances chemosensitivity. Genes Dev. 2011 Oct 15;25(20):2125-36. https://doi.org/10.1101/gad.17276711.
48. Tononi, G., Sporns, O., & Edelman, G. M. Measures of degeneracy and redundancy in biological networks. Proceedings of the National Academy of Sciences, 1999;96(6), 3257-3262.
49. Edelman, G. M., & Gally, J. A. Degeneracy and complexity in biological systems. Proceedings of the National Academy of Sciences, 2001;98(24), 13763-13768.
50. Zhang B, Long Q, Wu S, Xu Q, Song S, Han L, Qian M, Ren X, Liu H, Jiang J, Guo J, Zhang X, Chang X, Fu Q, Lam EW, Campisi J, Kirkland JL, Sun Y. KDM4 Orchestrates Epigenomic Remodeling of Senescent Cells and Potentiates the Senescence-Associated Secretory Phenotype. Nat Aging. 2021 May;1(5):454-472. https://doi.org/10.1038/s43587-021-00063-1.
51. Beck J, Turnquist C, Horikawa I, Harris C. Targeting cellular senescence in cancer and aging: roles of p53 and its isoforms. Carcinogenesis. 2020 Aug 12;41(8):1017-1029. https://doi.org/10.1093/carcin/bgaa071.
52. Liu B, Zhou Z, Jin Y, Lu J, Feng D, Peng R, Sun H, Mu X, Li C, Chen Y. Hepatic stellate cell activation and senescence induced by intrahepatic microbiota disturbances drive progression of liver cirrhosis toward hepatocellular carcinoma. J Immunother Cancer. 2022 Jan;10(1):e003069. https://doi.org/10.1136/jitc-2021-003069.
53. Guan Y, Zhang C, Lyu G, Huang X, Zhang X, Zhuang T, Jia L, Zhang L, Zhang C, Li C, Tao W. Senescence-activated enhancer landscape orchestrates the senescence-associated secretory phenotype in murine fibroblasts. Nucleic Acids Res. 2020 Nov 4;48(19):10909-10923. https://doi.org/10.1093/nar/gkaa858.
54. Leon KE, Buj R, Lesko E, Dahl ES, Chen CW, Tangudu NK, Imamura-Kawasawa Y, Kossenkov AV, Hobbs RP, Aird KM. DOT1L modulates the senescence-associated secretory phenotype through epigenetic regulation of IL1A. J Cell Biol. 2021 Aug 2;220(8):e202008101. https://doi.org/10.1083/jcb.202008101.
55. Loo TM, Miyata K, Tanaka Y, Takahashi A. Cellular senescence and senescence-associated secretory phenotype via the cGAS-STING signaling pathway in cancer. Cancer Sci. 2020 Feb;111(2):304-311. https://doi.org/10.1111/cas.14266.
56. Schmitz CRR, Maurmann RM, Guma FTCR, Bauer ME, Barbé-Tuana FM. cGAS-STING pathway as a potential trigger of immunosenescence and inflammaging. Front Immunol. 2023 Feb 28;14:1132653. https://doi.org/10.3389/fimmu.2023.1132653.
57. Sun L, Wu J, Du F, Chen X, Chen ZJ. Cyclic GMP-AMP synthase is a cytosolic DNA sensor that activates the type I interferon pathway. Science. 2013 Feb 15;339(6121):786-91. https://doi.org/10.1126/science.1232458.
58. Ishikawa H, Ma Z, Barber GN. STING regulates intracellular DNA-mediated, type I interferon-dependent innate immunity. Nature. 2009 Oct 8;461(7265):788-92. https://doi.org/10.1038/nature08476.
59. Glück S, Guey B, Gulen MF, Wolter K, Kang TW, Schmacke NA, Bridgeman A, Rehwinkel J, Zender L, Ablasser A. Innate immune sensing of cytosolic chromatin fragments through cGAS promotes senescence. Nat Cell Biol. 2017 Sep;19(9):1061-1070. https://doi.org/10.1038/ncb3586.
60. Li T, Chen ZJ. The cGAS-cGAMP-STING pathway connects DNA damage to inflammation, senescence, and cancer. J Exp Med. 2018 May 7;215(5):1287-1299. https://doi.org/10.1084/jem.20180139.
61. Fenech M, Kirsch-Volders M, Natarajan AT, Surralles J, Crott JW, Parry J, Norppa H, Eastmond DA, Tucker JD, Thomas P. Molecular mechanisms of micronucleus, nucleoplasmic bridge and nuclear bud formation in mammalian and human cells. Mutagenesis. 2011 Jan;26(1):125-32. https://doi.org/10.1093/mutage/geq052.

62. Takahashi A, Loo TM, Okada R, Kamachi F, Watanabe Y, Wakita M, Watanabe S, Kawamoto S, Miyata K, Barber GN, Ohtani N, Hara E. Downregulation of cytoplasmic DNases is implicated in cytoplasmic DNA accumulation and SASP in senescent cells. Nat Commun. 2018 Mar 28;9(1):1249. https://doi.org/10.1038/s41467-018-03555-8.
63. Takahashi A, Okada R, Nagao K, Kawamata Y, Hanyu A, Yoshimoto S, Takasugi M, Watanabe S, Kanemaki MT, Obuse C, Hara E. Exosomes maintain cellular homeostasis by excreting harmful DNA from cells. Nat Commun. 2017 May 16;8:15287. https://doi.org/10.1038/ncomms15287.
64. Zhang L, Pitcher LE, Yousefzadeh MJ, Niedernhofer LJ, Robbins PD, Zhu Y. Cellular senescence: a key therapeutic target in aging and diseases J Clin Invest. 2022 Aug 1;132(15):e158450. https://doi.org/10.1172/JCI158450.
65. Sun Y, Wang X, Liu T, Zhu X, Pan X. The multifaceted role of the SASP in atherosclerosis: from mechanisms to therapeutic opportunities. Cell Biosci. 2022 May 31;12(1):74. https://doi.org/10.1186/s13578-022-00815-5.
66. Coryell PR, Diekman BO, Loeser RF. Mechanisms and therapeutic implications of cellular senescence in osteoarthritis. Nat Rev Rheumatol. 2021 Jan;17(1):47-57. https://doi.org/10.1038/s41584-020-00533-7.
67. Kowald A, Passos JF, Kirkwood TBL. On the evolution of cellular senescence. Aging Cell. 2020 Dec;19(12):e13270. https://doi.org/10.1111/acel.13270.
68. Mitteldorf J. What Is Antagonistic Pleiotropy? Biochemistry (Mosc). 2019 Dec;84(12):1458-1468. https://doi.org/10.1134/S0006297919120058.
69. Giaimo S, d'Adda di Fagagna F. Is cellular senescence an example of antagonistic pleiotropy? Aging Cell. 2012 Jun;11(3):378–83. https://doi.org/10.1111/j.1474-9726.2012.00807.x.
70. Bonafè M, Valensin S, Gianni W, Marigliano V, Franceschi C. The unexpected contribution of immunosenescence to the leveling off of cancer incidence and mortality in the oldest old. Crit Rev Oncol Hematol. 2001 Sep;39(3):227-33. https://doi.org/10.1016/s1040-8428(01)00168-8.
71. Vasto S, Carruba G, Lio D, Colonna-Romano G, Di Bona D, Candore G, Caruso C. Inflammation, ageing and cancer. Mech Ageing Dev. 2009 Jan-Feb;130(1–2):40–5. https://doi.org/10.1016/j.mad.2008.06.003.
72. Arai Y, Martin-Ruiz CM, Takayama M, Abe Y, Takebayashi T, Koyasu S, Suematsu M, Hirose N, von Zglinicki T. Inflammation, But Not Telomere Length, Predicts Successful Ageing at Extreme Old Age: A Longitudinal Study of Semi-supercentenarians. EBioMedicine. 2015 Jul 29;2(10):1549-58. https://doi.org/10.1016/j.ebiom.2015.07.029.
73. Rolland Y, Sierra F, Ferrucci L, Barzilai N, De Cabo R, Mannick J, Oliva A, Evans W, Angioni D, De Souto Barreto P, Raffin J, Vellas B, Kirkland JL; G.C.T-TF group. Challenges in developing Geroscience trials. Nat Commun. 2023 Aug 19;14(1):5038. https://doi.org/10.1038/s41467-023-39786-7.
74. López-Otín C, Blasco MA, Partridge L, Serrano M, Kroemer G. Hallmarks of aging: An expanding universe. Cell. 2023 Jan 19;186(2):243-278. https://doi.org/10.1016/j.cell.2022.11.001.
75. Zhu Y, Tchkonia T, Pirtskhalava T, Gower AC, Ding H, Giorgadze N, Palmer AK, Ikeno Y, Hubbard GB, Lenburg M, O'Hara SP, LaRusso NF, Miller JD, Roos CM, Verzosa GC, LeBrasseur NK, Wren JD, Farr JN, Khosla S, Stout MB, McGowan SJ, Fuhrmann-Stroissnigg H, Gurkar AU, Zhao J, Colangelo D, Dorronsoro A, Ling YY, Barghouthy AS, Navarro DC, Sano T, Robbins PD, Niedernhofer LJ, Kirkland JL. The Achilles' heel of senescent cells: from transcriptome to senolytic drugs. Aging Cell. 2015 Aug;14(4):644-58. https://doi.org/10.1111/acel.12344. Epub 2015 Apr 22.
76. Grootaert MOJ, Moulis M, Roth L, Martinet W, Vindis C, Bennett MR, De Meyer GRY. Vascular smooth muscle cell death, autophagy and senescence in atherosclerosis. Cardiovasc Res. 2018 Mar 15;114(4):622-634. https://doi.org/10.1093/cvr/cvy007.
77. Chinetti-Gbaguidi G, Colin S, Staels B. Macrophage subsets in atherosclerosis. Nat Rev Cardiol. 2015 Jan;12(1):10-7. https://doi.org/10.1038/nrcardio.2014.173.

78. Zhou H, Khan D, Hussain SM, Gerdes N, Hagenbeck C, Rana M, Cornelius JF, Muhammad S. Colchicine prevents oxidative stress-induced endothelial cell senescence via blocking NF-κB and MAPKs: implications in vascular diseases. J Inflamm (Lond). 2023 Nov 24;20(1):41. https://doi.org/10.1186/s12950-023-00366-7.
79. Vellasamy DM, Lee SJ, Goh KW, Goh BH, Tang YQ, Ming LC, Yap WH. Targeting Immune Senescence in Atherosclerosis. Int J Mol Sci. 2022 Oct 27;23(21):13059. https://doi.org/10.3390/ijms232113059.
80. Lynch M, Bucknall M, Jagger C, Wilkie R. Healthy working life expectancy at age 50 for people with and without osteoarthritis in local and national English populations. Sci Rep. 2022 Feb 14;12(1):2408. https://doi.org/10.1038/s41598-022-06490-3.
81. Pignolo RJ, Law SF, Chandra A. Bone Aging, Cellular Senescence, and Osteoporosis. JBMR Plus. 2021 Apr 2;5(4):e10488. https://doi.org/10.1002/jbm4.10488.
82. Holloway K, Neherin K, Dam KU, Zhang H. Cellular senescence and neurodegeneration. Hum Genet. 2023 Aug;142(8):1247-1262. https://doi.org/10.1007/s00439-023-02565-x.
83. Chinta SJ, Woods G, Rane A, Demaria M, Campisi J, Andersen JK. Cellular senescence and the aging brain. Exp Gerontol. 2015 Aug;68:3-7. https://doi.org/10.1016/j.exger.2014.09.018.
84. Ungerleider K, Beck J, Lissa D, Turnquist C, Horikawa I, Harris BT, Harris CC. Astrocyte senescence and SASP in neurodegeneration: tau joins the loop. Cell Cycle. 2021 Apr;20(8):752-764. https://doi.org/10.1080/15384101.2021.1909260.
85. Limbad C, Oron TR, Alimirah F, Davalos AR, Tracy TE, Gan L, Desprez PY, Campisi J. Astrocyte senescence promotes glutamate toxicity in cortical neurons. PLoS One. 2020 Jan 16;15(1):e0227887. https://doi.org/10.1371/journal.pone.0227887.
86. Barbé-Tuana F, Funchal G, Schmitz CRR, Maurmann RM, Bauer ME. The interplay between immunosenescence and age-related diseases. Semin Immunopathol. 2020 Oct;42(5):545-557. https://doi.org/10.1007/s00281-020-00806-z.
87. Santoro A, Bientinesi E, Monti D. . Immunosenescence and inflammaging in the aging process: age-related diseases or longevity? Ageing Res Rev. 2021 Nov;71:101422. https://doi.org/10.1016/j.arr.2021.101422.
88. Pawelec G. Age and immunity: What is "immunosenescence"? Exp Gerontol. 2018 May;105:4-9. https://doi.org/10.1016/j.exger.2017.10.024.
89. Sharma R. Perspectives on the dynamic implications of cellular senescence and immunosenescence on macrophage aging biology. Biogerontology. 2021 Dec;22(6):571-587. https://doi.org/10.1007/s10522-021-09936-9.
90. Ong SM, Hadadi E, Dang TM, Yeap WH, Tan CT, Ng TP, Larbi A, Wong SC. The pro-inflammatory phenotype of the human non-classical monocyte subset is attributed to senescence. Cell Death Dis. 2018 Feb 15;9(3):266. https://doi.org/10.1038/s41419-018-0327-1.
91. Hall BM, Balan V, Gleiberman AS, Strom E, Krasnov P, Virtuoso LP, Rydkina E, Vujcic S, Balan K, Gitlin II, Leonova KI, Consiglio CR, Gollnick SO, Chernova OB, Gudkov AV. p16(Ink4a) and senescence-associated β-galactosidase can be induced in macrophages as part of a reversible response to physiological stimuli. Aging (Albany NY). 2017 Aug 2;9(8):1867–1884. https://doi.org/10.18632/aging.101268.
92. Effros RB, Dagarag M, Spaulding C, Man J. The role of CD8+ T-cell replicative senescence in human aging. Immunol Rev. 2005 Jun;205:147-57. https://doi.org/10.1111/j.0105-2896.2005.00259.x.
93. Covre LP, De Maeyer RPH, Gomes DCO, Akbar AN. The role of senescent T cells in immunopathology. Aging Cell. 2020 Dec;19(12):e13272. https://doi.org/10.1111/acel.13272.
94. Larbi A, Fulop T. From "truly naïve" to "exhausted senescent" T cells: when markers predict functionality. Cytometry A. 2014 Jan;85(1):25-35. https://doi.org/10.1002/cyto.a.22351.
95. Zhao Y, Shao Q, Peng G. Exhaustion and senescence: two crucial dysfunctional states of T cells in the tumor microenvironment. Cell Mol Immunol. 2020 Jan;17(1):27-35. https://doi.org/10.1038/s41423-019-0344-8.
96. Minato N, Hattori M, Hamazaki Y. Physiology and pathology of T-cell aging. Int Immunol. 2020 Apr 12;32(4):223-231. https://doi.org/10.1093/intimm/dxaa006.

97. Huff WX, Kwon JH, Henriquez M, Fetcko K, Dey M. The Evolving Role of CD8+CD28- Immunosenescent T Cells in Cancer Immunology. Int J Mol Sci. 2019 Jun 8;20(11):2810. https://doi.org/10.3390/ijms20112810.
98. Liu Z, Liang Q, Ren Y, Guo C, Ge X, Wang L, Cheng Q, Luo P, Zhang Y, Han X. Immunosenescence: molecular mechanisms and diseases. Signal Transduct Target Ther. 2023 May 13;8(1):200. https://doi.org/10.1038/s41392-023-01451-2.
99. Chou JP, Effros RB. T cell replicative senescence in human aging. Curr Pharm Des. 2013;19(9):1680-98. https://doi.org/10.2174/138161213805219711.
100. Laphanuwat P, Gomes DCO, Akbar AN. Senescent T cells: Beneficial and detrimental roles. Immunol Rev. 2023 Jul;316(1):160-175. https://doi.org/10.1111/imr.13206.
101. Elyahu Y, Hekselman I, Eizenberg-Magar I, Berner O, Strominger I, Schiller M, Mittal K, Nemirovsky A, Eremenko E, Vital A, Simonovsky E, Chalifa-Caspi V, Friedman N, Yeger-Lotem E, Monsonego A. . Aging promotes reorganization of the CD4 T cell landscape toward extreme regulatory and effector phenotypes. Sci Adv. 2019 Aug 21;5(8):eaaw8330. https://doi.org/10.1126/sciadv.aaw8330.
102. Fulop T, Larbi A, Pawelec G, Khalil A, Cohen AA, Hirokawa K, Witkowski JM, Franceschi C. Immunology of Aging: the Birth of Inflammaging. Clin Rev Allergy Immunol. 2023 Apr;64(2):109-122. https://doi.org/10.1007/s12016-021-08899-6.
103. Zhu H, Blake S, Kusuma FK, Pearson RB, Kang J, Chan KT. Oncogene-induced senescence: From biology to therapy. Mech Ageing Dev. 2020 Apr;187:111229. https://doi.org/10.1016/j.mad.2020.111229.
104. Yaswen P, Campisi J. Oncogene-induced senescence pathways weave an intricate tapestry. Cell. 2007 Jan 26;128(2):233-4. https://doi.org/10.1016/j.cell.2007.01.005.
105. Graziano S, Gonzalo S. Mechanisms of oncogene-induced genomic instability. Biophys Chem. 2017 Jun;225:49-57. https://doi.org/10.1016/j.bpc.2016.11.008.
106. Barnoud T, Schmidt ML, Donninger H, Clark GJ. The role of the NORE1A tumor suppressor in Oncogene-Induced Senescence. Cancer Lett. 2017 Aug 1;400:30-36. https://doi.org/10.1016/j.canlet.2017.04.030.
107. Peeper DS, Shvarts A, Brummelkamp T, Douma S, Koh EY, Daley GQ, Bernards R. A functional screen identifies hDRIL1 as an oncogene that rescues RAS-induced senescence. Nat Cell Biol. 2002 Feb;4(2):148-53. https://doi.org/10.1038/ncb742.
108. Angelini PD, Zacarias Fluck MF, Pedersen K, Parra-Palau JL, Guiu M, Bernadó Morales C, Vicario R, Luque-García A, Navalpotro NP, Giralt J, Canals F, Gomis RR, Tabernero J, Baselga J, Villanueva J, Arribas J. Constitutive HER2 signaling promotes breast cancer metastasis through cellular senescence. Cancer Res. 2013 Jan 1;73(1):450-8. https://doi.org/10.1158/0008-5472.CAN-12-2301.
109. Courtois-Cox S, Genther Williams SM, Reczek EE, Johnson BW, McGillicuddy LT, Johannessen CM, Hollstein PE, MacCollin M, Cichowski K. A negative feedback signaling network underlies oncogene-induced senescence. Cancer Cell. 2006 Dec;10(6):459-72. https://doi.org/10.1016/j.ccr.2006.10.003.
110. Ohtani N, Zebedee Z, Huot TJ, Stinson JA, Sugimoto M, Ohashi Y, Sharrocks AD, Peters G, Hara E. Opposing effects of Ets and Id proteins on p16INK4a expression during cellular senescence. Nature. 2001 Feb 22;409(6823):1067-70. https://doi.org/10.1038/35059131.
111. Faget DV, Ren Q, Stewart SA. Unmasking senescence: context-dependent effects of SASP in cancer. Nat Rev Cancer. 2019 Aug;19(8):439-453. https://doi.org/10.1038/s41568-019-0156-2.
112. Lau L, David G. Pro- and anti-tumorigenic functions of the senescence-associated secretory phenotype. Expert Opin Ther Targets. 2019 Dec;23(12):1041-1051. https://doi.org/10.1080/14728222.2019.1565658.
113. Acosta JC, Banito A, Wuestefeld T, Georgilis A, Janich P, Morton JP, Athineos D, Kang TW, Lasitschka F, Andrulis M, Pascual G, Morris KJ, Khan S, Jin H, Dharmalingam G, Snijders AP, Carroll T, Capper D, Pritchard C, Inman GJ, Longerich T, Sansom OJ, Benitah SA, Zender L, Gil J. A complex secretory program orchestrated by the inflammasome controls paracrine senescence. Nat Cell Biol. 2013 Aug;15(8):978-90. https://doi.org/10.1038/ncb2784.

114. 115. Yin K, Patten D, Gough S, de Barros Gonçalves S, Chan A, Olan I, Cassidy L, Poblocka M, Zhu H, Lun A, Schuijs M, Young A, Martinez-Jimenez C, Halim TYF, Shetty S, Narita M, Hoare M. Senescence-induced endothelial phenotypes underpin immune-mediated senescence surveillance. Genes Dev. 2022 May 1;36(9-10):533-549. https://doi.org/10.1101/gad.349585.122.
115. 116. Hashimoto H, Güngör D, Krickeberg N, Schmitt J, Doll L, Schmidt M, Schleicher S, Criado-Moronati E, Schilbach K. TH1 cytokines induce senescence in AML. Leuk Res. 2022 Jun;117:106842. https://doi.org/10.1016/j.leukres.2022.106842.
116. 117. Lee S, Schmitt CA, Reimann M. The Myc/macrophage tango: oncogene-induced senescence, Myc style. Semin Cancer Biol. 2011 Dec;21(6):377-84. https://doi.org/10.1016/j.semcancer.2011.10.002.
117. 118. Ovadya Y, Landsberger T, Leins H, Vadai E, Gal H, Biran A, Yosef R, Sagiv A, Agrawal A, Shapira A, Windheim J, Tsoory M, Schirmbeck R, Amit I, Geiger H, Krizhanovsky V. Impaired immune surveillance accelerates accumulation of senescent cells and aging. Nat Commun. 2018 Dec 21;9(1):5435. https://doi.org/10.1038/s41467-018-07825-3.
118. 119. Takasugi M, Yoshida Y, Ohtani N. Cellular senescence and the tumour microenvironment. Mol Oncol. 2022 Sep;16(18):3333-3351. https://doi.org/10.1002/1878-0261.13268.
119. Simova J, Sapega O, Imrichova T, Stepanek I, Kyjacova L, Mikyskova R, Indrova M, Bieblova J, Bubenik J, Bartek J, Hodny Z, Reinis M. Tumor growth accelerated by chemotherapy-induced senescent cells is suppressed by treatment with IL-12 producing cellular vaccines. Oncotarget. 2016 Aug 23;7(34):54952–54964. https://doi.org/10.18632/oncotarget.10712.
120. 121. Salminen A, Kauppinen A, Kaarniranta K. Myeloid-derived suppressor cells (MDSC): an important partner in cellular/tissue senescence. Biogerontology. 2018 Oct;19(5):325-339. https://doi.org/10.1007/s10522-018-9762-8.
121. 122. Wang T, Notta F, Navab R, Joseph J, Ibrahimov E, Xu J, Zhu CQ, Borgida A, Gallinger S, Tsao MS. Senescent Carcinoma-Associated Fibroblasts Upregulate IL8 to Enhance Prometastatic Phenotypes. Mol Cancer Res. 2017 Jan;15(1):3-14. https://doi.org/10.1158/1541-7786.MCR-16-0192.
122. 124. Kong SH, Ma L, Yuan Q, Liu X, Han Y, Xiang W, Liu DX, Zhang Y, Lu J. Inhibition of EZH2 alleviates SAHA-induced senescence-associated secretion phenotype in small cell lung cancer cells. Cell Death Discov. 2023 Aug 5;9(1):289. https://doi.org/10.1038/s41420-023-01591-y.
123. 125. Chen C, Xu P. Cellular functions of cGAS-STING signaling. Trends Cell Biol. 2023 Aug;33(8):630-648. https://doi.org/10.1016/j.tcb.2022.11.001.
124. 126. Prata LGPL, Ovsyannikova IG, Tchkonia T, Kirkland JL. Senescent cell clearance by the immune system: Emerging therapeutic opportunities. Semin Immunol. 2018 Dec;40:101275. https://doi.org/10.1016/j.smim.2019.04.003.
125. 127. Antonangeli F, Zingoni A, Soriani A, Santoni A. Senescent cells: Living or dying is a matter of NK cells. J Leukoc Biol. 2019 Jun;105(6):1275-1283. https://doi.org/10.1002/JLB.MR0718-299R.
126. 128. Song P, An J, Zou MH. Immune Clearance of Senescent Cells to Combat Ageing and Chronic Diseases. Cells. 2020 Mar 10;9(3):671. https://doi.org/10.3390/cells9030671.
127. 130. Frey B, Derer A, Scheithauer H, Wunderlich R, Fietkau R, Gaipl US. . Cancer Cell Death-Inducing Radiotherapy: Impact on Local Tumour Control, Tumour Cell Proliferation and Induction of Systemic Anti-tumour Immunity. Adv Exp Med Biol. 2016;930:151-72. https://doi.org/10.1007/978-3-319-39406-0_7.
128. 131. Toso A, Di Mitri D, Alimonti A. Enhancing chemotherapy efficacy by reprogramming the senescence-associated secretory phenotype of prostate tumors: A way to reactivate the antitumor immunity. Oncoimmunology. 2015 Jan 22;4(3):e994380. https://doi.org/10.4161/2162402X.2014.994380.
129. 132. Toso A, Revandkar A, Di Mitri D, Guccini I, Proietti M, Sarti M, Pinton S, Zhang J, Kalathur M, Civenni G, Jarrossay D, Montani E, Marini C, Garcia-Escudero R, Scanziani E, Grassi F, Pandolfi PP, Catapano CV, Alimonti A. Enhancing chemotherapy efficacy in Pten-deficient prostate tumors by activating the senescence-associated antitumor immunity. Cell Rep. 2014 Oct 9;9(1):75-89. https://doi.org/10.1016/j.celrep.2014.08.044.

130. 133. Laberge RM, Sun Y, Orjalo AV, Patil CK, Freund A, Zhou L, Curran SC, Davalos AR, Wilson-Edell KA, Liu S, Limbad C, Demaria M, Li P, Hubbard GB, Ikeno Y, Javors M, Desprez PY, Benz CC, Kapahi P, Nelson PS, Campisi J. MTOR regulates the pro-tumorigenic senescence-associated secretory phenotype by promoting IL1A translation. Nat Cell Biol. 2015 Aug;17(8):1049-61. https://doi.org/10.1038/ncb3195.
131. 134. Chang J, Wang Y, Shao L, Laberge RM, Demaria M, Campisi J, Janakiraman K, Sharpless NE, Ding S, Feng W, Luo Y, Wang X, Aykin-Burns N, Krager K, Ponnappan U, Hauer-Jensen M, Meng A, Zhou D. Clearance of senescent cells by ABT263 rejuvenates aged hematopoietic stem cells in mice. Nat Med. 2016 Jan;22(1):78-83. https://doi.org/10.1038/nm.4010.
132. 135. Islam MT, Tuday E, Allen S, Kim J, Trott DW, Holland WL, Donato AJ, Lesniewski LA. Senolytic drugs, dasatinib and quercetin, attenuate adipose tissue inflammation, and ameliorate metabolic function in old age. Aging Cell. 2023 Feb;22(2):e13767. https://doi.org/10.1111/acel.13767.
133. Fuhrmann-Stroissnigg H, Santiago FE, Grassi D, Ling Y, Niedernhofer LJ, Robbins PD. SA-β-Galactosidase-Based Screening Assay for the Identification of Senotherapeutic Drugs. J Vis Exp. 2019 Jun 28;(148). https://doi.org/10.3791/58133.

MIX
Papier aus verantwortungsvollen Quellen
Paper from responsible sources
FSC® C105338

If you have any concerns about our products,
you can contact us on
ProductSafety@springernature.com

In case Publisher is established outside the EU,
the EU authorized representative is:
Springer Nature Customer Service Center GmbH
Europaplatz 3, 69115 Heidelberg, Germany

Printed by Libri Plureos GmbH
in Hamburg, Germany